STOP THE "FUELISHNESS"

STOP THE "FUELISHNESS"

*Plan For A World W/o Fossil Fuels
Save The Environment*

Johnroy Messick, PE

Copyright © 2015 by Johnroy Messick, PE

Library of Congress Control Number:	2015919898
ISBN: Hardcover	978-1-5144-3107-8
Softcover	978-1-5144-3108-5
eBook	978-1-5144-3106-1

All rights reserved. No part of this book may be reproduced or transmitted in any form or by any means, electronic or mechanical, including photocopying, recording, or by any information storage and retrieval system, without permission in writing from the copyright owner.

Any people depicted in stock imagery provided by Thinkstock are models, and such images are being used for illustrative purposes only.
Certain stock imagery © Thinkstock.

Print information available on the last page.

Rev. date: 01/21/2016

To order additional copies of this book, contact:
Xlibris
1-888-795-4274
www.Xlibris.com
Orders@Xlibris.com
540628

Contents

0	Opening Statement	1
00	Pre-foreword	10
000	Foreword	17
1.0	Start of Research	28
2.0	What Happens When We Run Out of FF?	30
3.0	Definitions	32
4.0	Abstract	38
5.0	Summary	40
5.1	Summary of Findings	41
5.2	Summary of Conclusions/Recommendations	48
5.3	Summary of World Plan*	54
6.0	Jobs	55
7.0	Introduction	60
8.0	Bio-Background Continued	63
9.0	World Supplies of Fossil Fuels (FF) Resources	65
9.1	World supplies of FF (Ej Btu)	66
10.0	World Energy Consumption and Projections*	68
10	Energy Consumption and Projections (Shell Oil)	70
11	Power Generation Cost (All Fuels)	72
11.1	Food to Fuel	73
11.2	Solar	76
11.3	Wind Power	79
11.4	Fossil Fuels.*	84
11.5	Nuclear	87
11.6	Fusion	88
11.7	Summary	90
11.8	Hydropower	91
12.0	How Do Nuclear Fuels Fit into This Picture?	93
13	How Do We Plan for a Sustainable and Cheap Energy Future?	97

14	Nuclear Power Around the World	100
14.1	Great Britain	101
14.2	Germany	102
14.3	France	105
14.4	Argentina and Brazil	107
14.5	Russia	109
14.6	United States	113
14.7	China*	115
14.8	India*	117
14.9	South Korea	119
14.10	Japan	120
14.11	Iran*	123
14.12	Hungary	125
14.13	Africa	126
14.14	Turkey	128
15.0	Closed and Open Nuclear Power Systems	129
16.0	The Tail-end of Nuclear Power: Nuclear Spent Fuel	130
17	Updated Global Energy Outlook, 2014	132
18	Smaller Power Plants	133
19	Nuclear Power - Environmental Issues	137
20.0	Nuclear Power Safety Issues	164
21.0	Philosophy of Life	169
21.1	Ayn Rand	170
21.2	Johnroy Messick: The Renaissance man	171
22	Economics	176
23.0	Pandora's Promise	190
24	Closing Statement	197

This book was written from a U.S. perspective. Research funded by Johnroy, not Big Oil or Koch.

Notice the cover picture-- Straight out of Compton (South Central LA). This book is straight out of Gardena CA-- neighbors of Compton and Watts. And it contains a triple entendre for our book.

0

Opening Statement

John R. Messick, 4/15/15

Truth – Honesty – Knowledge

To solve and stop the "fuelishness" and to save the planet.

Our research has led us to envision work that

1. Eliminates all carbon emissions and saves our oxygen.
2. Saves millions of lives from fossil fuel emissions and avoids 1.25 million deaths, worldwide from traffic accidents.
3. Reduces power generation cost by 50% compared to FF.
4. Reduces fuel cost by 99.44% (e.g., a fuel bill of $400/mo becomes less than $4/mo). This will require a brand-new transportation system that's connected directly to the electricity grid.
5. Extends supplies of crude oil from 30 years to 10,000 years—today. I'm a successful licensed professional engineer and project manager in five states.
6. I volunteer to relocate to China, if necessary, to manage its huge projects. China already has many huge projects: the largest airport, the largest dam (Three Gorges), and the longest underwater tunnel (under construction). China values engineers and experts. The United States has many elements that don't.
7. Four countries are mostly already on the risk path to success. The United States is not one of these countries yet. That's it, folks! All the evidence is herewith.
8. After discovering this knowledge, you will come to believe what I believe: Energy from atoms is the people's energy, and it takes a collective of humans to convert the matter into the "people's power." Oligarchs or autocrats cannot do *this*. So let's get to work.

Final Word

The world consumes 10,000 billion megawatts per hour of power dumping 3,600 billion tons of CO_2 into the atmosphere. Precious oxygen *needed* to *sustain* flora and fauna are being *consumed* at a rate of 1,200 and 2,400 billion tons per 100 years, respectively. Prior to the Industrial Revolution, these rates were 99% less. By visualizing this continuing for another 1,000 years, one can begin to see that a sustainable fauna life maybe in trouble. This applies to many species, including humans. This situation can be predicted via Newton's second law of thermodynamics, that is, the disorder of the world goes to infinity[*] unless humans step in and take control over what they have created.

Civilizations exist because humans harvest energy and use it to sustain an otherwise difficult life. At the same time though, humans must deal with the shit[**] this creates. Or like other aspects of life, disorder (entropy) gets out of control, and the endpoint is life that is unsustainable all in accordance with the laws of thermodynamics (heat, energy, and matter). Regarding this matter, it's not the numbers that matter; it's the matter that matters. And sometimes even matter never matters.

An example is a conversation I was having with my adult daughter. I have a young daughter also.

Daughter: Was it tough being a parent of a bipolar daughter?

Me: Sometimes, but the toughest thing I ever had to do was to tell you, as a young girl, that you were adopted.

Daughter: But it didn't matter.

Me: "It didn't matter because it never mattered to you or to your mom and me."

One final thought on entropy—the human species cause most of the environmental entropy (shit), and only humans can reduce entropy.

And now the new transportation system is getting personal. Yesterday, January 21, 2015, I was run over by a hit-and-run driver. I was at the corner of Wilshire Boulevard and Roberson in Beverly Hills

---

[*] Infinity can be either heaven or hell. (Our Lord left this up to us.)

[**] Shit, either solid or gaseous, and we know it's harder to capture gas. But it's all shit.

and on my way to visit my daughter. I'm still in the hospital today (23). This is the second time I have been run over since my stroke. I don't drive so I have to walk or ride the bus. You will read more about my condition herewith. I will let you know in Chapter 24, if I survive. This got me thinking about our individual responsibilities as a member of the world's civilization. Just like this situation, the responsible thing to do is own up to our involvement and pay the consequences. Don't run away and hide, convincing ourselves that the responsibility is not ours. Only criminals respond this way.

The same can be said regarding anticlimate emissions. The responsible thing for the world to do is to access the damage already caused and have all countries contribute to a fund to (1.0) pay the victims and (2.0) pay for the clean-up. This fund should be funded by fees accessed from new emissions and fees accessed from emissions caused during the production of the assets of the world. This fund and its management should be added to the charter of the IPCC. This would demonstrate to the world the importance of this matter and the cost to clean up our mess. I didn't realize how handy my name would be for this assignment! Now onto the research.

<u>Separate Statement on Transportation</u>

I need to elaborate on transportation because it is a big reason why we are in the situation we find ourselves in today.

I came to this epiphany after recovering from the hit-and-run assault and having the time to read Mr. Caro's book* and reflecting on his research regarding Mr. Mose's lasting effect on New York City. I was guided to continue my research on how to use our knowledge on energy to correct our transportation system problems.

Consider this dichotomy vis-a-vis advancements and retractions.

First is oil consumption. The world consumes 100,000 barrels of oil per day (bpd). The United States consumes 20% of the total even though its population is only 5% of the whole. Other big consumers include China, utilizing 10%, Russia 5%, and India 4%. But worse than that is the United States imports 8,000 bpd and China 5,000 bpd, with Russia exporting 5,000 bpd, some to China and the United

---

* Robert A. Caro "The Power Broker - Robert Moses and the Fall of New York City"

States. So who will suffer the most when the oil crises hits sometime in the next 30 years?

Next is transportation. Most of the world's oil is consumed by our inefficient transportation system: mostly big private cars and small trucks. And it's only going to get worst as the world's wealth is spread more equally. This is a political issue with economists like Piketty (France) and Klugsman (United States) leading the way with their work and writings on blending capitalism and freedom values with socialist economics.

This transportation issue—the belief that private cars and trucks have the right to expand regardless of the consequences—needs to change. In Mr. Caro's book, he gets into the destruction caused by the projects of Robert Moses in New York City to accommodate the private car and truck issues today.

The epiphany came to me when Mr. Caro described how Moses, a.k.a. "the greatest public servant ever," became Moses, a.k.a. "the worst public servant ever." With Caro's seven years of research and 1,400 pages of writings, he chronologizes with meticulous adherence to the facts surrounding Mr. Moses's brilliance with bullying and using his influence to get his way with weak-kneed politicians of his time. It took the strength of politicians such as Franklin Roosevelt and Nelson Rockefeller to rein in Moses's misapplied vision and to move forward with mass transit for New York City, a city that is still suffering from these projects of Moses's era.

And what was the root cause of Moses's failings? His lack of compassion and humility had led him to building mostly for the needs of Fat Cats and their big private cars, including destroying communities and the environment with concrete, steel, and asphalt to accommodate these Fat Cats getting out of town fast in their big cars. Moses never realized the dichotomy he was facing with regard to big private cars and the ultimate destruction they were causing. His cynosure was his lust for power. Mr. Caro describes how private cars in New York City became a tyranny of tragic accidents and jams. If this can happen to a great city like New York and if the world uses these techniques to solve traffic problems, then the entire world will be paved over, and all life will be suffering with air that has 800 ppm CO_2, causing extreme droughts, floods, and storms.

As one last research effort, I looked into how many humans are killed and injured each year due to vehicular accidents. I couldn't believe what I discovered.

1.24 Million Humans Are Killed Each Year

Four hundred thousand of these people were either pedestrians or bicyclists without a lethal weapon to fight with (source: World Health Organization).

Comment: If we can't live in a society where it's safe for animals, pedestrians, and bicyclists, then we need to engineer a better society.

Back to Moses. Moses expected his political power would lead to greatness. He never came to the realization that political power never leads to greatness and wisdom. Only the people's power can achieve greatness and wisdom.

Mr. Caro developed his compassion from all his research on the most compassionate president of the United States, Lyndon Baines Johnson.

The good news is that we have two options to stop this fuelishness:

One, tether all big trucks, trains, and buses and ban all other transportation vehicles except golf carts in communities for those that can't pedal a bicycle.

Two, tether all transportation vehicles to the electrical grid powered by 2C electricity, except carts.

The country that develops the most efficient nonpolluting systems will have enough resources left over to help other areas of the world. This help will lead to greater "people power" across the world. This is an equal sum game, with the power that oligarchs have with $25 fossil fuel power. The people will have two cents of nuclear power. If the United States wants China and Russia to occupy this space, then stay on the path we are on, the United States can lead the world to occupy this space, and hence our motto "A risk path going forward is a better choice than a safe path going backward."

Some will ask why not batteries? There are four reasons:
1. Batteries are expensive.
2. Batteries are heavy (some weigh more than the passenger load).
3. Batteries waste other valuable resources.
4. They are inefficient by up to 25% for small cars and other battery-operated vehicles. These vehicles have to carry the batteries with

them wherever they go. Why do all vehicle manufacturers spend lots of money to reduce the vehicle's weight? They do this to increase the miles traveled per million BTU.

Read our final statements in chapter 24.

This new transportation system will be an <u>engineered</u> system with the following criteria:

1. Tether the vehicles to an electrical grid powered with 2C fuel.
2. Design to prevent accidents that kill.
3. Use computerized routing and backup systems.
4. Eliminate all anticlimate emissions.

Consider this: The system we have today was never <u>engineered</u>. It came about based on using technology as it became available. Our forefathers sponsored the national freeway system primarily to solve the mortality rate caused by two-lane traffic <u>not</u> separated with a median strip. So now it's our time to solve the problems of today.

The roads built by our forefathers will be fully utilized in the engineered system. The evolution keeps on going, going, and going.

The new engineered transportation system will require a new generation of small cars with three new primary features, namely, an extension cord, one electric motor, and a tiny battery. The small battery will be used to propel the car a mile or two between electrical connections. Keep in mind that these are just the criteria. The system and car will require engineering to formalize and prove the concepts either with scientific calculations or with prototypes before gearing up to full production and implementation.

Google can work with us to implement some of its concepts for a smart car with these features. Engineers and scientists go through this "proof of concept" phase before proceeding with plans and specifications.

The benefits are as follows:

1. Avoid many of the transportation deaths.
2. Save 99.44% of transportation fuel cost.
3. Save the environment.
4. Solve many of the traffic jams.

a. More people will be using public transportation.
b. Libertarians would rather not drive than to give up some freedoms.
c. Drunks will not be able to connect into the complicated system. And when they do they will immediately be disconnected when they go off course.

The Road to "The Pursuit of Happiness"

The writing of this book has helped me understand one route to happiness. I know many of you are on this road.

The eternal drive for knowledge (the truth) will push us over the hills of misery into the valleys of happiness, and we don't stop there. For those of us who have the luxury of time and/or the knowledge and inheritance to amass a fortune, we owe a debt to those who are working to support us.

Happiness comes when this debt is being paid off. If a debt remains upon our death, it's up to the "state" to settle the debt. This frees up resources so that many more will have an equal opportunity for "the pursuit of happiness."

Final Word

Read Chapter 14: China's nuclear technology will be the best, utilizing the best from Russia and the West. Russia is cornering the uranium market by purchasing the uranium supplies of the world, including 20% of North American supplies.

00

Pre-foreword

Be reminded that top governments of the world invented the nuclear bomb and delivery systems that were used to end a very long war. In the same way, we are in a race with the world to develop nuclear power, IFR, fusion, and waste technologies for future generations. And unless it's owned by the world, governments, and world companies such as Terra, GE, WH, and Shaw Engineering, we could be held hostage for the cheap 2C power.

A little story: I authored a technology book on motor fuel alcohol (gasohol), a DOE contract renewable fuel via corn 2.5 gal/bu. I worked for RKII, an engineering/construction company. The Chinese contacted us, and I hosted a delegation from China in1978. The Chinese wanted to purchase our technology. It was already in the book, except engineering. The delegation included a professor, a plant manager, a government official, and a security person. We sat down together and began the discussion. Each person had a copy of my DOE book (250 pages). They knew the book better than I did.

They got mad when we would not allow taking pictures of the plants. One of these plants was a reactor company in St. Louis. The only thing they didn't understand when they left town was—how could I be the driver when I'm supposed to be an engineer? We never heard from them again. Sometimes, one learns a lot just by experience and observations—the heuristic method. Enjoy the book. I already have. I guess I'm a solipsistic at heart.

Another story about engineers:

GE Edison engineers developed a processing system that would manufacture 10,000 light bulbs per minute. The startup was painful, which led to GE selling all the engineers to Belcan Co. Later, other engineers got the process working. And now GE's manufacturing cost are the lowest in the industry.

The reason GE sold the engineers was because it was an economic issue; GE could hire the engineers back without these engineers participating in GE's profit-sharing program. There were many engineers making big-bucks. Belcan is a temp service and engineering company. I was VP technology. Slavery is still alive and well in the West. All temp-service companies participate in this slavery program.

So maybe there's hope for Obamacare 12/4/12. As an engineer/project manager, I can assure you the bigger the project, the bigger the problems. It's the rule of unintended consequences (Murphy's Law) at work.

Successful progress toward accomplishing big projects is always slow. But the public thinks things are accomplished via Moor's Law—technology advances fast, not Murphy's. And politicians get confused as to which law applies.

Another short story:

I once owned a 1971 Toyota Land Cruiser, and the parts' costs were high. Looking around, I discovered I could go to a junkyard and pick up parts that fit from a 1954 GMC truck. And who wanted GMC to go bankrupt? Certainly not the Japanese. If we don't value our engineers/scientist, then others will. No smarts, just looking around.

We have been sharing our technology/engineering all these years and didn't even know it. So why not do our part with the international community and finish the work for 2C power for the whole world to use. If we don't and others do, we could be held hostage for cheap power.

Read about Russia and China doubling nuclear power with 30 to 50 new plants by 2015 (Chapter 14), Russia funding approximately one-half trillion dollars for research on fusion/fast cycle reactors, and China building most of the new nuclear power plants under construction today, using Westinghouse technology.

And what do you think FF 4c power will cost when it starts paying for carbon emissions? Maybe 5–6c power compared to the 2C of nuclear power! Big Oil is already considering the power cost increase to cover the carbon emission levies. *(New York Times, 12/5/13. Read on.)* The knowledge here explains the many reasons why FF consumption can be minimized before we run out or FF gets very expensive.

Speaking of knowledge, Jeremy Black has a good book on the subject regarding *Power of Knowledge,* delving not only on the importance of knowledge to plan our future but also on the nexus and synergies between information, military, economic, political and social factors. Francis Bacon was an early writer on knowledge, with his writings in the book, *Knowledge Is Power.* Bacon explores the scientific revolution back to René Descartes and how this use of knowledge was used to shape the distinctiveness of the West from the eighteenth to the twenty-first century.

Black, on the other hand, explores the syllogism between data and knowledge. Many of our politicians today don't even want to acknowledge the data available to the world nowadays. And the data today is a revolution in itself, with the advent of computers and programmers. These are beyond the scientific revolution of Bacon's day. Climate change and its nexus with anticlimate gases are a good example, but not the only one. Nuclear power and its benefits for the world is another. There are many examples of science that are being ignored. We have been electing antiscience politicians like_____? You can fill in the blank.

You will find in my Philosophy of Life (Chapter 21) where I accidentally stumbled on the three tenets of wisdom and how some philosophies of life will never lead to wisdom because these philosophies are self-centered and will prevent one from achieving the last stage (third tenet) of wisdom. OK, I will tell you now, third stage: compassion with passion.

And here's the bottom line, for me: If I were to have profits from this book, I will probably give these away. I have already committed to give 25% to the church on Adams St. in Los Angeles.

One final point: Let's redistribute monies on a pro rata scale and redistribute energy to the world via nuclear projects. So there, Norquest and Ayn. Take that to the bank.

As Mandela stated "from dust to dust," we say from poor as a child back to poor as a senior. What a ride! I would not have it any other way, except a little less inequality.

It was Eisenhower who said, "Atoms for peace," and to which we add, "Atoms to power out of poverty."

The DCF-ROI analysis used to support electricity generation cost shows up in Chapter 11. Now what is the important question you need to know? Because many use this variable to answer any tendentious question (e.g., if I want the price of power to be 1 cent/kWh), all I have to do is input the variable that gives this to me. As I explain herein, I wrote a computer program for this analysis and used sensitivity analyses on this variable.

So I know how to manipulate the numbers. You will have to read on to answer the question.

Note to the reader: "I used this question to get to our label *2C power for the world*, this book's leitmotif. And then expanded it to 2C for the environment.

Hint: In today's low-interest-rate environment, 2C is achievable. And remember, FDR said that if you weren't born then, "the only thing we have to fear is fear itself."

So without this fear, I was able to write this book. Chapters 19, 20, and 21 will provide anecdotes of knowledge to help arrest your fears.

Breaking News

After completing our research and most of our book, I read Mr. Bryce's new book on energy. There appears to be a contradiction between us and Mr. Bryce. He doesn't think FF are being depleted fast. All other conclusions are the same. Mr. Bryce claims that the iron law of economics will balance supply and demand.

However, he stops short and doesn't explain what this means vis á vis prices. This is where we can help. We will put numbers on the matter. This leads us to realize there are *no* differences between Mr. Bryce and us. All of us believe the FF are being depleted fast, such as the DOE, United States, and Shell, with three caveats:

1. Relatively stable prices
2. No major technology advances
3. Consumption levels that support a reasonable growth

Regarding technology, breakthroughs with IFR, thorium, and spent waste recycling for nuclear power will extend supplies from 25,000 years to 200,000+ years. We explain that this will not occur without dispelling all the fear and lies the EV are espousing, before we can build out our world plan.

Read about how these lies and fears are financed and funded by Big Oil and King Coal.

Regarding the first caveat, a short story first then onto the numbers and cost. Back in the 1960s, I purchased gasoline for my Nash Metropolitan (first four-cylinder) for *26 cents per gallon.*

With gasoline at $5.00 per gallon today, our hypothesis is that if the same 2,000% increase in energy prices occur over the next 50 years in order to bring supply and demand into balance, the prices will be:

$100 per gallon of gasoline and
$100 per million BTU

This represents a 2,000% increase in these next 50 years.

And *nuclear* will never be more than $1 dollar.

Read-on to discover more about how to go nuclear for transportation.

Discount what I say later in the book: "With this matter, it's not the numbers that matter it's the matter that matters." I was referring to all the numbers in the research. Now you will see that I use you, we, us, and our because you are already a partner in our world collective of angels. Good night, Ayn. Welcome aboard! Enjoy.

Our label is 2C energy for the world and the environment.

Our motto—A choice between a risk path going forward—is better than a safe path going backward.

A short story about my Nash convertible, or we will never get out of here!

I purchased the Nash convertible for $60.00 (a little extra dollars I had then) I could fill up for $2.00. Once I was driving from Auburn to get to my coop job at St. Regis Paper in Pensacola, Florida. I was running out of gas (empty). I came upon this filling station (we called them filling stations then). I didn't have any money, so I charmed the attendant (a black) into lending me the gasoline until I passed back through. He was one of my angels along the risk path.

You know the rest. For $2.00, I could travel the roundtrip (about 200 miles). This cost about one cent per mile. A new car today costs approximately $25,000.00, and the fill-up costs $100.00, so the cost per mile might be twenty-five cents (no calculator) for the same 2,000% increase. When you get to nuclear transportation, you will discover what we and the DOE mean and how to get on that path. One more fact: Tuition at Auburn was $50.00 per quarter (back to the same 2,000% increase).

And if my angels had not been placed there, I would have been stuck in the middle of Alabama. I would have survived although, because I am the Renaissance Man . . . *free, white, and over twenty-one.*

Actually, I was twenty at that time. So the numbers, cost, and prices demonstrate how Bryce and us have no differences. With these prices, many more people will not be able to purchase FF energy. The only difference is that FF will only be available for the "one-percenters."

At these energy prices, how many of our descendants will not be able to afford power, much less gasoline? Even today, two billion of our brethren do not have electricity. In the 1960s, it was only 500 million. This is the real meaning of supply and demand. Not just words. The title of this book could just as well been: *Planning for a World with Very Expensive Fossil Fuels.*

A good example of this inequality shows up in Gardendale, Texas (the heart of the Eagle Ford aquifer), where there is plenty of expensive FF, but the residents live in extreme poverty. The area produces $15 million worth of oil every day. The residents live in makeshift communities called *colonias*. (*New York Times*, 6/30/14)

Lyndon Johnson taught school there. His experiences in Gardendale inspired him to take on poverty as president. Both LBJ and MLK were driven more by poverty than race. Their major issue was the extreme inequalities that creates a caste system where 39% of children live in extreme poverty. Thirty-nine percent is today's number. As FF become more expensive, many more communities will become *colonias*. Does it matter whether or not we deplete the last gallon of FF or whether the last gallons cost $100.00 per gallon? No answer needed.

000

Foreword

I dedicate this book to my three children and granddaughter—Meg and John are in their thirties, and Eden and Charlotte are both eleven (2011). Without their love and my second wife (Joyce), I would not have survived the massive hemorrhage stroke I suffered in 2006, a nigh death experience. That's why I'm so pleased to offer my knowledge and guidance on this matter, a culmination of my life's oeuvre. I plan to spend the rest of my life supporting this matter.

Let's all plan so that future generations can enjoy a wonderful life with plenty of cheap energy. If we start today, I'm convinced this can be achieved, but time is of the essence!

Come along and take this journey with me. I hope you have some fun. How could I not, with the opportunity to share my life's oeuvre, that is, my knowledge, education, and experience. Applying heuristic truths, we can seek to

1. save so many lives,
2. save the earth's atmosphere,
3. provide cheap power for future generations, and
4. point out the weaknesses with Western democracies before these weaknesses destroy the freedom and justices enjoyed by all of us that live in the West today.

Book Mission Statement

To develop a framework for a plan to extend world energy resources for at least 25,000 years, with the dual secondary benefits of:
1. minimizing climate-destroying pollution, and
2. improving the world's economies.

The United States, with Mr. Obama, is in the best position to lead this effort, even though the country is behind at least four other countries. Read on.

Conclusions/Recommendations and Plans

See Chapter 5. To the critics, this statement applies:

"I don't want to live in God's imperfect world because accidents (shit) happens."

Thanks to Harold Schultz for office space, Wi-Fi, food and drink at Magic Johnson's Starbuck's in Watts, California.

Thanks to Ford for shelter, and UPS for coping, etc.

Hypothesis

Like any good research project the hypothesis is this: Fossil fuels are finite and we are running out "fast." This will lead to an energy crisis soon, and we need to find cheap alternatives.

My Jesus knows I have done the best I can to contribute to the well-being for future generations.

One more: It's time to value engineers (PEs), scientists and experts.

They won't disappoint you. Just believe in them. We work hard. The best work of my career is for you and the plutocrats.

I remember as a 7-year-old boy, before my dad dug a well, stealing the tractor while my parents were away. I drove down to the local spring to fetch two buckets of clean water. I also remember topping the hill, seeing my parents. They got home before I did. I knew my dad was going to beat me for stealing the tractor. It was my job to fetch the water. I was lucky the tractor didn't run out of gasoline. My parents scolded me but didn't beat me. My brothers and sisters were happy. I'm the oldest of six. (My mother was 26.) One problem I had was how to balance the two five-gallon cans while driving. It was an easy problem to solve.

I also remember my mother being so happy when she heard about a government program funded by taxpayers (i.e., the rural electric program sponsored during the Truman presidency). The power company would provide free poles if the farmer would install the poles. See, even then the government was considering taxpayers. We eventually moved to another worn-out farm when I was 11, one that had electricity for the well pump.

I remember my mother crying on the stairs about ten steps from the top of the staircase. She was clutching the new vacuum cleaner my granddad had purchased. She was so happy, and I was so happy for my mother. My granddad inspired my life's journey. My mother still had to hand-wash our clothes on a ribbed plank, and we never had running water. Somehow, we ended up with a washing machine when I was 12. Again I was so happy for my mother. Then when I was 13, my mother took all of us children, and we escaped from our dad and went to live in the city with my mother's parents.

If we had not escaped, I'm convinced that our father would have killed all of us. A side story: When I was eight, it was Steve, who was six years old, and I who had to take care of the cows while my dad was away from the farm working (driving a truck). He got home early one morning and decided to go out to the barn to check on the cows. Well, it just so happened that one of the cows became bloated, a common problem caused by the cow eating too much grass with wild onions. My dad took me out to the barn, and we began to work on the cow by running a hose through its ass to get to the stomach. The cow was struggling. It was time for me to walk to the school bus stop. As I was leaving the barn, my father turned to me and said, "If this cow dies, I will kill you when you come home." Of course, I had a terrible day at school. Not knowing what to do, I remember thinking about running away and not returning home. My decision was not to run away. My father was so mean I knew if I ran away, he would find me and kill me. I decided to take my chances with the cow.

Since I'm here, as Paul Harvey would say, "Here's the rest of the story."

Sorry for the side trip. Let's get back on our path. It's now 4:00 a.m., 11/22/11.

I'm going to direct the editor to put these last two pages in just as I wrote them. I don't have time to correct these pages because on Monday, I will have to talk with my editor (whoever that might be) to see if I should go to the Philippines to direct the data input. I'm accustomed to having talented secretaries that can decipher my hen-scratching and misspelled words. If needed, I will drive my F-150 Ford van (home) to the Philippines and stay for the time it takes for the data input (schedule and budget for three to four months).

Now I promise to go back to bed. If you don't appreciate the excursion, too bad. I already have your book money, and it's my money and time. I never had the guts to tell my boss before—I don't give a shit what you think. Like it or not. Now surcease. It's 10:30 a.m.

Next time, I will tell you about walking one mile up the muddy rut of a road to meet my mother and help her carry home my newborn brother, Charlie, in a cardboard box from the hospital at two in the morning. I was then 10 years old. To this day, I don't know how I found out she would be coming home that day. No other adults were in the house.

We didn't have a phone. How can you have a phone without electricity? Now it's 11:30 a.m.

I was inspired to tell those stories last night so that you have a personal story about how hard life can be without electricity, even in the richest country in the world. So our Lord is telling all of us to help the other 2 billion people without electricity.

Now, where do I put this! I have broken all the rules on how to write a book and will be breaking all the rules on writing a research project. Please excuse me. My teachers would be mad at me.

OK, we made it, 12/1/13. What a trip. I thought the Philippines was close to Hawaii. But wasn't that fun. We almost ran out of gas. It's a good thing we were in a F-150 Ford Van instead of a 350. Reminds me of the time I flew my plane (Moony) to Puerto Rico. We were working on a water project. Fortunately, I didn't have my family with me. Anyway, I stopped in the Bahamas. The place was out of gas. So I went on to Grand Isle. After arriving, I discovered they would not have gas for another three days. Well. (I got that from President Reagan.)

Back then, I was always in a hurry. I looked at my maps and found an airport with gas in the Dominican Republic. The problem was that I only had twenty minutes of gas left. I figured it would take me nineteen minutes to get there. Just like John Kennedy Jr., I took the risk path. I figured if I had to crash, at least it would be in water. I would try to get close to shore, away from the sharks. Back to the story: The F-150 Ford Van with big balloon tires and a bladder. Allen Mulally was smart to think of building an amphibious van to handle the business once the oceans rise another twenty feet. Like Noah, he must have been inspired.

The island is in bad shape because of a typhoon, but where I am, at least the buildings are still standing. The Filipinos are in good spirits.

The Philippines is beautiful, and so are the ladies. Even with the destruction, the high spirit is everywhere. If you need something to inspire you on being passionate about anticlimate emissions, just visit here. The people are working hard (at least, they have jobs) to put things back together with the big help of the U.S. government, led by Obama, Boehner, and Reid.

Ralph Nader should come here to witness the effects of such a storm. Then maybe he would not be so mendacious with his nuclear rhetoric. Curcease! My editor just told me this book is supposed to be nonfiction (boy, she is pretty). Nader is the leader of the environmental villains, EVs, that are against nuclear power (2C power). Can you even imagine EV being on the side as Big Oil and King Coal. This is what fear, money, and power can do to a person. Read on and revisit the CNN program, *Pandora's Promise*. The program is summarized in Chapter 23.

I like the Farmers Insurance advertisement:
"The more you know, the better you
can plan for what's ahead."

This research discredits the activists EV malicious information on the price of electricity for nuclear power versus other renewable fuel sources, in addition to other important matters.

Another good book by Rodney Stark is *How the West Won* "because of political freedom and scientific knowledge." He surveys the great empires of the past, including the Egyptian, Roman, Ottoman, and Chinese. Then he goes on to discuss what factors led to their downfall. One is greed of the ruling class, with the masses stuck in poverty and with the elites viewing work as degrading. This is a sign of the elitists' superiority and how the rulers killed the use of technology—technology threatened their domain. The elites consumed rather than invested, while indulging themselves at every turn. These empires declined not so much from things they did, but rather because others ascended to Western values and capitalism, while at the same time having the right mix of freedom, property rights, and education plus the willingness to use reason to advance knowledge. "The industrial Revolution was born."

Now let's use this knowledge to realize that the West could lose it all if we are fearful to utilize knowledge to go forward! And there are signs that this is already happening for elite monarchies and oligarchies: (1) work is degrading; (2) climate change is not happening; (3) experts are to be feared; (4) and voter suppression has already started to take away the freedoms of democracies. Read about the Ayn's cults (Chapter 21) of the world having no use for collectives. The first thing to go is the rights of all these working people. As you can see, I'm a member of the Solipsistic Hegelian Society where everyone is welcomed. Just leave your guns at the door.

Our ancestors knew that when the sun is shining bright in the summer, one needs to plan for winter when the sun is not shining. Now we need to discover that we can't leave our planning to the FF industries and Big Oil. Their incentive is to beat next quarter's numbers. The only party that can do this planning for our society is a big government via our Constitution. Big G has the responsibility for the general welfare of the country. Let's ask the tea party who else can do this for the world? So far, we have *not* been smart enough to elect leaders that can figure this out. It just takes an engineer who has the time and retirement money to fund this task. I hope the politicians are at least smart enough to understand what's in this book. I don't have much hope though. Many can't even add 2 + 2 = 4 regarding climate change issues and LED bulbs. At least, Jimmy Carter knew this much.

The above discussion was prepared before I started my research on nuclear fuel resources and supplies. Now, after the Japan Fukushima disaster, I didn't expect much. However, the result was far different. I am now convinced that nuclear power is the only means for the world to have sustainable cheap 2C power and save the environment for future generations. This is quite a change for an engineer that started as a Jane Fonda antinuke. Other energy resources have a place, but must be managed around nuclear power. What I mean here is for all of us to conserve our limited resources and to use only within a managed approach to sustainable energy. The primary reasons for this conclusion is:

1. Nuclear power resources are about limitless at over 250,000 years compared to oil's only 30 years.
2. To achieve no. 1 above, we will need to continue our research and engineering on the following:
   a. Recycling of spent fuel materials
   b. Light and fast reactor (IFR) program
   c. Thorium technology
   d. Fusion

   Without these new technologies, we have nuclear resources for 25,000 years. But with tomorrow's technology, its limitless.

3. Utilize the Fukushima accident to engineer solutions around massive earthquakes and tsunamis.

These engineering, science, and technology advances will not happen overnight. Educational and funding needs to be acknowledged and accepted by politicians and the public. We already pay the DOE and others, such as the Congressional Resource Research council (CRR) to collect the data, but our politicians don't use the information to educate the public. That's why we need more engineers as politicians to teach us what the problem will be if we continue on our current path. Now the push back on this might be that when we had an engineer as our president (Jimmy Carter), he was viewed by some as ineffective. However, he sponsored growth in nuclear science and started many conservation measures. And he funded coal-to-oil projects. President Carter was a nuclear engineer without industry experience and he had a Energy Secretary that was anti nuclear. In addition, congressional politicians did not understand what he was trying to achieve. Consider this, no matter how the present data are extrapolated, the end of fossil fuel (FF) resources is within our sight and can never be replaced.

Every pound we burn is gone forever (except as carbon dioxide). President Carter knew this also.

Our politicians can focus on the national debt because this is something they (think) know about. However, the debt crisis will be a picnic compared to the looming expensive energy crisis. Can you imagine how difficult it will be to educate our current politicians when many of our leaders today can't even understand why LED lighting is important and how items like this contribute to a viable future?

Can you imagine discussions by future generations about how our generation wasted their future's viability? Also consider this: the Chinese and the Russians may have an edge over Western civilizations.
1. They plan for longer time periods.
2. Their policies don't have to be accepted by the public.
3. They are ahead of the Western world except France and India with their nuclear programs.
4. They can plan their society based on the research of experts.

Recently, China has proven how successful this approach can be, advancing from the bottom to the second greatest economy. I believe we can live up to the challenge facing us, but we must vote out some of our leaders. A part of the plan is to start a planning process that considers our energy consumption and supplies for at least 50 years. This can bring the public into the process. The good news is that once fossil fuels are consumed, CO_2 levels will at least stabilize at 800 ppm. Today, we are at 400 ppm, up from 250 ppm before the Industrial Revolution.

Our contribution to this is to recommend the United Nations sponsor an Energy Sustainability Council (UNESC).

Can we survive at 800 ppm? The climate scientists report that we are already suffering at 400 ppm, and they advise to start reducing emissions, so that the temperature rise can be limited to 2C.

After watching CNN's TV special program *Pandora's Promise* on 11/7/13, I woke up at 2:00 a.m. and started this addition to the foreword. The TV show was so powerful I knew I had to get my book published as rapidly as possible. So, I will stop the research I was working on the Ayn Rand philosophy and with errors that haven't been flushed out or other sections that could use some more work.

Pandora's Promise came to the same conclusions as this book does regarding nuclear power as an answer for our environment. The show came to this conclusion based on the environmental and health issues. With our research, we shall get there based on a plan to provide cheap power for at least 25,000 years; otherwise, we'll run out of cheap oil in 30 years. The world needs a cheap source of energy. The French nuclear program is a model the world could use, going from 0% to 80% nuclear in only 20 years. See Chapter 14 for a discussion on "Nuclear Power Around the World."

Now I can go back to bed; it's 4:00 a.m.

How can the nuclear matter happen when the EV and FF industries are on the same side? They have a lot of power. But once the public is on the nuclear side, then the world collective will have more power than Big Oil and King Coal! Post *Pandora's Promise;* we have much more evidence. the Philippine tragedy is one example. Looks like it's getting worse every day. So after the public hears about this book, the world collective will have the truthful facts and will be on nuclear's side. Al Gore is getting his revenge for having the Presidency stolen from him. But he certainly isn't happy.

Our world climate problems are more reasons the world needs big governments to tackle these big global problems. So if Ayn Rand and her cult doesn't like collectives, they will have a hard time sleeping at night: once this matter happens.

To wrap up this foreword, here's one additional point—No company or government could issue a report like this:

1. No company because Big Oil and King Coal have a lot of money and power, and these companies don't want to lose any business.
2. No government because the governments are afraid of the EV and EA.

So without a job, I hired myself for the job.

Just this morning, 11/25/13, the *New York Times*, in p. A5, reported: "Mercer Germany in trouble. Need to further increase energy prices." This would not be funny, if it wasn't such a self-inflicted wound. Sounds familiar? If nothing else, maybe this book will keep us from making the German mistake, just in case some crazies like Ralph get into office. I promise this is my last comment. (Don't go to the bank yet.) Actually, I thought I was in Chapter 25. So keep going forward. You will read my prediction about Germany later on in Chapter 14.

So now that we have plowed our way through the forewords, let's get started with the research at 2:00 a.m.

After competing all the research, I almost had to throw the forewords away and start over. But since I have my soul in the above, I decided to move-on. I will give you this important information. At the end of the research, after all the knowledge and satisfying the project mission and hypothesis, we came up with two diamond-in-the-rough gems. These gems are worth the project cost. These gems lead to the biggest project ever. You can find these gems summarized in Chapter 5.

1.0

Start of Research

The first task was to research FF resources/supplies worldwide and then determine how long these supplies will last if the world consumes FF at the current and projected consumption rates. This research and analyses require a lot of technical data and information. So the best way to read this book is through a cursory review of the data and assumptions; realize I have done my homework and am knowledgeable about the subject. If you don't want to know about my background or philosophy, skip these chapters. On the other hand, you may want to read everything because of the narrative as the project progressed. Also, a benefit is my humor (not grammar).

The second task was to work up the generation cost for all power fuels. Don't be disappointed because I didn't do a DCFROI analysis on all fuels. I did extract DCFROI data from the EIA, including nuclear (to establish the 2C label). Other pricings were mostly established by surveying the market for pricing information (e.g., NH wind over land or ocean plus Germany's experience with wind energy). FF pricing is mostly available. One point on wind. The very first two power resources for the world were wind and solar. One would have thought the energy harvesting cost for these two sources would be competitive with FF. So maybe there's a reason (Mr. Icahn). Even private industry isn't smart all the time. I can't leave out governments. Just look at what's happening to Germany. Alternatively, France is the example for all of us.

You will observe here with the abbreviated folio the materials used for references. With Google and Bing, we don't need the arcane method regarding detailed references listed in the back of the book. Many of us never liked it anyway. Now, references can be imbedded in the palimpsest. This method allows for the timeline and references to be a part of the story.

If your government system doesn't have a voting public, just be reminded our collective will be judging you on the five Es. I only became a raging, progressive liberal after my research showed that the plutocrats and the Ayn Rand cult, mostly conservatives, are gaining a hold on media outlets via either not reporting news events or worst reporting misrepresentations and omissions of the facts. The WSJ, Fox News and Rush Limbaugh's Radio Station are examples of this in the United States. I'm sure other countries in the world have their problems with the media, more specifically, China, Russia, and former Soviet Union countries. If it's available at all. See Chapters 14, 19, and 24.

2.0

What Happens When We Run Out of FF?

When I started my research for this book, I thought I would write an apocalypse, something like the book Apocalypse 2012. But since we survived 2012, we see that such (written in 2011) predictions become only scare tactics to sell books. With this research, I have formulated a plan to prevent a major catastrophe. The best way to approach this is to start with a statement made by the DOE in 2010:

Without timely mitigation, world supply/demand balance will be achieved through massive demand destruction (shortages), accompanied by high price increases, both of which could create a long period of significant economic hardship worldwide waiting until world conventional oil production peaks before initiating crash program mitigation leaves the world with a significant liquid fuel deficit for two decades or longer.

Our statement: The United States consumes more liquid fuel (oil) then the total of the other top consuming countries, such as, China, Russia, and Germany.

Bryce's new energy book disagrees with our hypothesis. The difference is that Bryce believes that higher prices will force a rebalancing of supplies. We and the DOE disagree with Bryce, and here's the reason: "These higher prices will force many (more) of our brethren out of the FF marketplace via not being able to purchase the energy in the future." We already have two billion people without access to electricity.

I believe this is an accurate framing of the subject. For oil, the problem is that our politicians do not want to face up to this reality. Plus, they have no energy knowledge. So it's easy to latch on to information that is palpable to the public, in other words, making the issue politically acceptable for a scenario. How do we expect anything much different when the path to becoming a politician does not include the scientific community, experts, or the poor?

A 1970 prediction by the DOE was that oil/gas and coal would last for 50 and 600 years, respectively. But this prediction didn't address all FF. One reason for this is that several years ago, China, India, and other countries were not purchasing U.S. FF. The DOE forecasts were based only on our resources not the world supply and consumption situation we find ourselves involved with today. The current projection of thirty and 200 years is based on 2% world growth. If the growth becomes 4%, then the projection might be 15 and 100 years. Again, no matter the actual projections, these can be debatable. What's not debatable by the scientific community is that FF are finite, whereas nuclear resources are almost infinite. (See Chapter 12)

Just today in the *New York Times*, 4/11/11, it was reported that Congress and the President passed legislation to cut funding for the nuclear recycling project called "the Shaw Arena MOX project located at the Savannah river nuclear complex in South Carolina." Read Chapters 19 and 20 about the cuts in funding for Yucca Mountain. Many believe those funding cuts will kill the projects and, therefore, the energy needs of our children and future generations. We have already spent over $10 billion developing the knowledge. The Clinton administration started the projects after considering methods to handle the huge inventory of spent fuel nuclear materials. These projects alone could convert 44 tons into a useable nuclear fuel, enough to power 43 large reactors for a year, 10% of the world's nuclear reactors. More important, the science and engineering would be advanced to recycle the spent fuel materials from all reactors. Without these advances, nuclear energy technology will be set back for many years. This is the knowledge our generation owes to future generations. The "thorium" knowledge is the work (jobs) for future generations.

After watching *Pandora's Promise*, the news that night was that the Obama administration canceled funding for the IFR reactor program. Read Chapters 19, 20, and 12.

3.0

Definitions

As we delve into this technical matter. Just know, it's not the numbers that count, it's the matter that counts.

Btu = 1059.67 joules £1.0 Rj
0002931 kWh = Btu
252 cal = Btu
hp = .7355 kW
Btu = the amount of energy needed to heat one pound of water one degree F
1 calorie = 4.184 Joule
0.25 kcal (large) food
K = 1000
M = million
kWh = kilowatt-hour
W = watt
h = hour
g = gram
1 quadrillion Btu × 10^{15} = 1.0e × A (j) joule × 10^{18} joules
1MM Btu = $(28.26 \text{ meters})^3$ of x of natural gas
1 barrel oil = 42 gal. @ 8.4165/lb/gal.
1 lb oil or coal ~ 12,000 Btu/lb. (Read the reason below why coal is a problem to define.
1 lb of Uranium = 35,000,000,000 Btu
1 quad fuel = produces 11 gigawatt years
MM = million
Mega = million

Other useful information:
1 quad of energy from:
- 1 trillion cubic feet CH4
- 45 million tons coal

-15 million lbs Uranium
- 170 million barrels of oil
- electricity = 33 gigawatt hours
a typical power plant takes 3× fuel input for 2× energy output.
per capita energy use – 1,819 kg/oil equivalent/day
1 barrel oil = 5.63 g joules
1 million cubic meters CH4 = 34,700 g joules
1 kWh = 3.6M j × 10 ^6
giga = 1-^9
tera = 10^9 (trillion)
q = quadrillion = 10^{15}
World population: 8.5 billion people rounded to 10 billion
1 gram U = to 3 tons coal (energy)
MW = megawatt
FF = fossil fuels
NPP = nuclear power plant/s
2C environment = 2 degrees Celsius (label)
2C energy = 2 cents/kWh (label)
IEA = International Energy Agency
EA = environmental activist/s
EV = environmental villain/s
EE = energy efficiency
WSJ = *Wall Street Journal*
NYT = *New York Times*
LAT = *Los Angeles Times*
ITER = International Thermonuclear Reactor
IPCC = International Panel on Climate Change.
IAEA = International Atomic Energy Agency.
UNESC = United Nations Energy Sustainability Council (new - tomorrow)
LA = Los Angeles

Limitless + 1000 generations
Fission/Fusion – sometimes used interchangeably, incorrectly. This book should not be used as a resource for nuclear energy.

Conservationist (Example)

I taught my children not to waste (i.e., the waste-not-want-not philosophy).

My son (MBA Xavier) lives in Northern Kentucky. During one winter day, he got a telephone call from Duke Power.

Lady: Are you stealing electricity?

Son: No, I just don't use much.

Lady: How can you explain why your usage is less than what it takes to run a refrigerator?

Son: When one keeps the house cold, the refrigerator doesn't run much.

Silence.

He never heard from Duke again.

I'm a proud dad.

EE and conservation are components of our world plan. NYT, 10/9/14, in "The Problem With EE" reads: The advancement of LEDs this year, with the Nobel Prize in physics going to three researchers who discovered how to radically increase the EE of LED lighting, resulted in more energy consumed because of cheaper illumination. In fact, the IEA and IPCC reported that these efficiencies could backfire by 50% globally. Now our plan with cheap 2C power is exactly what we want: cheap power for all the world, especially the poor. Even with 2C power, if LEDs can reduce lighting cost by 50%, this essentially means 1C cost for illumination with no anticlimate emissions.

And later, we learn that when we stop burning our food supplies, the direct savings are $7 billion per year because of tax subsidies to Big Oil. The money ends up going to corn growers, with higher corn prices.

The indirect cost to consumers is over $1trillion dollars. (See Chapter 11.1)

Just this month (4/14) the company Opower went public. Its business model is to work with power companies to help save energy usage for the company's customers, says Alex Laskey, one of the cofounders of the company located in Arlington, Virginia. Mr. Laskey explains Opower uses behavioral economics along with statistics to prove how much money can be saved. Opower claims that 20 million customers have saved 4 terawatt hours of electricity and will soon be saving this per year the equivalent of the electricity generated by the Hoover Dam. This is how important energy conservation can be to the world. So why is the Ayn Cult so against LED lighting? By the way, this amount of savings is worth $10 billion per year for 4c power or $5 billion for 2C power.

Example - How to Calculate the Size of a Power Plant

The average household uses 1 kw/h.,-a 1600-megawatt power plant will service 1,600,000 homes.

So if the electric bill is $50.00/mo, and the price is 5 cents/kw, this household uses $50.00/$.05 = 20 × 100 = 1000/30 × 24 = about 1.3 kw.

Another calculation (without a calculator):

The average price for electricity in the United States is 11 cents/kWh. A lot of places have electricity at a price of 4 cents/kwh. The bill is $20.00/mo. Sometimes, power/electricity and energy are used incorrectly, like the power company when the electricity company is technically correct.

CO_2 Emissions from Oil, Coal, and Gas

The popular belief is that natural gas will save 50% of the CO_2 expelled when burned compared with coal.

The actual numbers are 900/for CH_4 per xs power and 1,300 per xs power for coal.

For a savings of about 40% (source: my memory, no calculator).

For even these lower reduced percentages there's a caveat—the 40% is based on older technologies for firing coal.

This information adds to the importance of nuclear power. If burning gas only saves 40% of the anticlimate gases compared to coal, nuclear saves 100% along with sun and wind. Renewables like food/shit expel almost the same as coal.

One reason this comparison is confusing is because coal is not:
1. hard coal (anthracite) releasing 29 to 33 kj/g
2. soft coal releasing 17 to 21 kj/g

So from one extreme to the other, the heat released doubles, and therefore, the same ratio goes for CO_2 emissions. We used the average for coal in the calculation above.

The media are too lazy to figure this out. So sci-fi journalists just write whatever is convenient or politically needed for an agenda. No wonder the public doesn't have a chance to understand the truth; more on this in the ensuing chapters.

When I use the information above and use updated efficiency numbers, burning CH_4 releases 35% CO_2 emissions than coal. Mostly, politicians and the media use the 50% number; some go as high as 70%. This helps EV sell the nuclear fears/lies.

hp - horsepower (use your sense perception)

Otherwise, 745.7 watts is almost the average home consumption. With regard to burning almost a ton of coal every 60 days, check me on that one. So let's get the horse out of our living room and the 300 horses out of the garage.

4.0

Abstract

- Fossil fuels (FF) resources are finite, and the world is consuming FF fast.
- Renewables "sun and wind" are free, but the harvesting cost is very expensive.
- Nuclear supplies are almost limitless, and the conversion cost to electricity is 2C, relatively cheap. In addition, nuclear power doesn't give off any anticlimate emissions.
- The things that are holding the world back are:
    1.0 geo/political systems and
    2.0 fear/lies spread by environmental villains (EVs) and financed by oligarchs.
- There are four nations that are further along "with the "world plan" than others. The United States is not one of these four.
- But the United States and others can catch up fast.
- I'm a disabled, retired chemical/petroleum engineer/project manager with Social Security Retirement Funds to support me while working this project. I have no fear because I have already lost the things that could be taken from me. This has allowed me, with the grace of my Jesus, to be totally free to write this book for the world. My biggest weakness is that I'm among the 47s club, and therefore, I'm lazy and a parasite to society. That's why it took so long to finish this project.

The journey's narrative is provided as a bonus.

Short Bio

John as a poor child was forced to work a worn-out farm without electricity, running water or telephone in Appalachia- by an abusive father. They dug ourselves out of this pit, when our dear Mother escaped with all six children. He then went on to finance 100% of his attendance to Auburn and Xavier, graduating with high honors, top engineering graduate; with memberships in the honor societies of Phi Kappa Phi- all top graduates, Tau Beta Pi- engineering and president of Phi Kappa Upsilon- chemistry. John worked 50 years as an Engineering Project Manager obtaining titles like President of the Ohio Association of Consulting Engineers, Youngest Project Manager and Technology-Vice President. He has two daughters, a son and a beautiful granddaughter; a second ex-wife that still loves me and a third that barely did.

Short Summary

John was inspired to write this about fuels-not fools- because of his suffering as a child without the conveniences of electricity, or fuel for heating, e.g., also running water or telephones; because all three require fuels. After 60 years studying/working as an Engineering Project Manager mostly on Energy/fuels; including projects like: coal to liquids and the Technology/Economics of Alcohol for Motor Fuels. On the latter he authored a book on the subject; both projects for the DOE/USA. He has had only one Government job- a county Sanitation Engineering position. He only lasted three months and quit after turning the water off on an Apartment Complex owned by the local Republican - He was threatened. At that time, He had two young children. The reason? The Owner by-passed the health permit regulations. Back to our book that started only about Energy, but morphed into the other Esque of life: ending with a plan to extend oils 30-25000 years, and saving the environment; while at the same time saving 99 & 44/100% of the worlds transportation fuel cost, and 50% of the cost of electricity generation. A bonus is the saving of 10s of thousands of deaths, and millions from pain and suffering with the Truth and only the Truth; with Knowledge and Compassion as the Props.

5.0

Summary

There are four countries that are already mostly on track to follow the recommendations and world plan outlined herein.
- France
- India
- China
- Russia

This conclusion was reached after completing Chapter 14. Iran will be on the list after completing the country's first project. The country has an ambitious nuclear program to not only supply its needs but also to export electricity to other countries in its region. Maybe Israel will purchase some of the electricity for the poor Palestinians that lost their power plant after the last war with Israel.

These countries still have much work to do to fully implement the world plan. These countries are mostly ahead of the rest of us.

Where is the great United States? We are not in the leading group, and the reasons are:
1. Our geopolitical system
2. The fear/lies spread by EV (environmental villains)

However, the United States and other Western nations will be able to catch up and lead in this matter once

1. this book is published,
2. the media and politicians become knowledgeable and start reporting the truth and only the truth,
3. the renaissance man and his world collective gets to work.

5.1

Summary of Findings

Power/fuel prices:

    Price per million Btu
    Oil - $25.00 recent history $5.0–25.00
    Gas - $3.50 recent history $2.0–16.00
    Coal - $3.50 recent history $ 1.0–3.50
    Nuclear - 2 cents recent history 2–10 cents
    Wind - free
    Sun - free
    Fusion - 30 years more research needed

Resources (supplies) at current consumption rates:

    Oil - 30 years
    Gas - 50 years (updated)
    Coal - 200 years
    Nuclear at "present technology" - 25,000 years. See why it may only be 1,000 years once the world goes 100% nuclear.
    Nuclear (completed technologies) - almost limitless (200,000+ years.)
    Wind and sun - limited. "Areas available to capture" at the generation cost are in Chapter 11
    Fusion - limitless once technology is complete

Generation cost (large plants) - Table 11.7-1,
Generation cost (smaller plants) - Table 18-1,
The average power price in the United States is 11cents/ kWh.
Once we convert our transportation from oil to nuclear, we go from $25/mm Btu to 2 cents/mm Btu (with zero emission). Stop dreaming!

Oil Consumption barrels/day (bpd):

Country	Consumption	Imports	Exports
United States	20,000	8,000	0
China	10,000	5,000	0
Russia	5,000	0	5,000
India	3,000	?	0
Japan	5,000	?	0
World Total	100,000		

Table 12.1

Sustainable Years of Nuclear Fuel Supply*

Potential energy production	Energy Potential energy production conventional tWh^18	Energy Potential energy production world nuclear electricity generation conventional (tWh^18)	Years Sustainable years at 1999 world nuclear electricity total conventional	Years Sustainable energy at 1999 world nuclear electricity generation (total)
Current fuel cycle	827,000	21,200,000	326	8,350**
only one recycle fuel cycle	930,000	23,900,000	366	9,410
Light water and fast reactor with recycle	1,240,000	31,800,000	488	12,500
fast reactor fuel cycle/recycle	+26,000,000	+630,000,000	10,000	25,000
Advanced thorium/uranium fuel cycle with recycling/fusion	++43,200,000	++90,200,000	Limitless	Limitless Rounded

*NEA Report 20.2 (Reference only – Added Recycle Technology (++).

**Rounded to 10,000 years

The 25,000 and limitless years are my projections which are very conservative.

CAVEAT

The 10,000 year projection is based on a 2%/yr growth. If the world accepts our plan, the growth in nuclear will be much more than 2%. Thats why the new energy sustainability council needs to be on top of this; and develop a graph charting: Sustainability Years vs: world conversion to nuclear. And provide energy direction for the world to use.

Table 11-7-1

Summary Generation Cost (Large Power Plants)*

Table 11-7.1 (from Chapter 11)

Fuel	cents/kWh
Nuclear	2.0 (rounded)
Wind	8.0 over land
Wind	17.0 over oceans (rounded)
Coal	3.0
Gas	3.0
Oil	**
Solar***	5.0 (rounded)

*Large power plants (1000+ megawatts)
**Oil is not competitive with coal or gas.

***The 5.0 cost is based on the projections from the California/Arizona plant. The most recent information is that the plant is having trouble at the 5.0 cost (data in Chapters 11 and 12)
Oil is used today to temper or for emergency needs (except in older plants).

Table 18-1

Summary Power Cost (Smaller Power Plants)

Fuel Comparisons for Smaller Plants

Fuel	Cost: cents/kWh
Nuclear*	2.0**
Wind	Add 100% to large plant cost
Coal	" " " "
Gas	" " " "
Oil	" " " "
Solar	" " " "

*See Chapter 18, Technology 90% Complete, Engineering and Prototype Development 20 to 61% Complete.

**The actual cost will be between 2 and 10 once the research/engineering is complete ~ 2025.

Note: Bryce, in his new energy book confirms the gas and nuclear numbers of 3 and 2, respectively. We got our cost numbers from the DCF ROI analysis provided by the DOE EIA.

Current world consumption (Ej):

1. Coal - 80
2. Oil - 140

3. Natural gas - 80
4. Nuclear - 55, this has already saved many lives and extended FF supplies

This means our oil-based economy cannot exist as we know it for more than 30 years. The price for FF will become very expensive. The hardships will begin sometime within these 30 years. So the most important thing we can do for our children and grandchildren is to plan for our entrance into this period. The matter of this book is to describe a plan for this reality. (Bryce agrees also)

A common theme among our politicians is to drill, drill, and drill or otherwise to quickly retrieve the resources available, and sometimes even to tap our oil reserves—this approach just makes the problem worse. These actions reduce prices, and we consume more of our FF supplies, and emit more and more anticlimate emissions and kill more and more people and other flora and fauna.

One thing the United States has accomplished over the years is to convince oil suppliers to trade paper for oil. We all know the problems this has created. It has benefitted the United States by supplying the oil we need for our economy. This certainly hasn't helped the world economy and the environment or prevented wars.

When I began my serious work on FF supplies in the early 1970s (Carter administration) with work on coal liquefaction, at that time, we had about 50 years of U.S. supply for oil. Then the world, mostly with our help, discovered the supplies of the Middle East, Alaska, and others, plus we added some nuclear power. So that's how we survived until now. Currently, the United States has discovered and pumped about 50% more FF today than in the 1960s. We are still importing 40% of our oil needs.

This brings up a point. Maybe we could do the same thing in the next 30 years. There are two major intervening factors and one benefit (engineering).

1. We have surveyed the world and discovered most of the easy resource supplies.
2. Back in the 1970s, the consumption of fossil fuels was only *150* Ej. With the expansion of democracy in countries like China, India, Indonesia, Brazil, Russia, and others, the consumption

has grown to about 300 Ej. With the projection of a 2% growth per year, the consumption will be 900 Ej by the end of the decade.

With the recent engineering for national gas and shale developments (see Chapter 17), the most this will do is extend FF supplies for 10 years. With our plan, supplies will be extended for 25,000 years; and with new technologies limitless.

Note: Bryce pointed out to us that supply and demand will take over, and FF will not be depleted this fast. Our research with "Bryce economics" will lead us to $100.00 per million Btu equivalent to $100.00 per gallon (our price projection) to balance supply and demand so that the last gallon is not consumed.

5.2

Summary of Conclusions/Recommendations

1. All new power plants should be nuclear.
2. Older FF plants should be replaced with nuclear or other renewables, as funding becomes available.
3. Initiate a program for the United States to lead a worldwide effort to plan around extending FF supplies for 25,000 years. This can be accomplished via the United Nations and other world agencies (see 11 below).
4. Organize all research on nuclear power via world organizations.
5. These world organizations can sanction the safest and most economical NPP throughout the world. This will prevent any country from keeping this technology and engineering from the rest of the world.
6. Companies within these countries can market these technologies, engineering, and prototypes to the world.
7. The combination of nuclear, knowledge, and absence of fear is very powerful. The world collectives have the power to eliminate all the monarchs and plutocrats of the world, wherever they exist.
8. The biggest nuclear accident in the United States was the Three-Mile Island plant in Pennsylvania. No one was killed or injured. The bigger accidents across the world are the Chernobyl and Fukushima accidents in the Ukraine and Japan, respectively.
   These accidents would not happen in the United States because of these reasons:
   8.1 Chernobyl - The United States has better reactor technology. A handful of people were killed. Not the millions the EV got paid to spread.
   8.2 Fukushima - the Japanese company, Keyota Electric, took shortcuts on the plant's investments. This would not happen in the United States because the country has better

regulations. Fukushima injured six workers but did not kill anyone. The tsunami killed 20,000.

9. Four weaknesses have appeared recently with Western democracy, namely:

    9.1 Media - Solve the problem of major news outlets reporting misinformation and omissions of important news or facts. This is so important because a successful democratic system depends on an educated public voting for candidates that support their view on the matter.

    9.2 Provide the funding (taxes) required for scientists/experts to research the matter, (e.g., UN climate panel, the Nuclear International Agency, etc.) so that the science and technology can be completed.

    9.3 Reinforce existing environmental agencies so that lies/fears spread by private industries and EV can be balanced with an independent government (big). Once a balanced position is available, the public can analyze both sides.

    9.4 Provide funding to educate our children and adults to a level that is consistent with the problems we currently have with the media and freedom of speech.

A fifth weakness should be added (e.g., the problem of an elite group of judges and lawyers that are isolated from the real world, leading to common folks not having access to the highest court available for justice). And even when access is achieved, the court, because of this isolation makes a ruling like the LA choke-hold case. These incompetent and isolated judges "set-back our evolution for centuries." Once the evolution-revolution occurs, this will be corrected.

These weaknesses are the same for other matters that are critical for the survival of democracy. Otherwise, the monarchs and plutocrats that believe in experts will control the power of the world.

10. Build a battery-powered bus system or a simulacrum therewith LA's metro system currently being developed using busses build by BYD Motors Inc., a Chinese company. Or better yet build an electric grid system to connect busses and big trucks directly to the electric grid. This will "practically" eliminate the need for cars or expensive batteries. For the few cars/trucks needed locally, these should be electrically operated.

10.1 This conclusion/recommendation is attached as an add-on to our recommendation to use the LA electrified bus system. If we still want cars and trucks, here is an engineering approach to the problem—a completely new engineered transportation system. Instead of stopping with a tethered bus system, what about a totally new transportation system without batteries?

Expand our thinking to envision a transportation system whereby all buses, cars and trucks are tethered directly to the electrical grid. Think of it simply at first: The tethered vehicle would be plugged into a socket; the socket would be on a raceway and the plug would be programmed to follow a route specific to that vehicle. The new-transportation system will not have the total flexibility to drive at speeds of 100 mph and will not be able to leave the programmed path to crash into other vehicles. There's a huge benefit for a new system like this: Our existing system with untethered vehicles kills 1.25 million people worldwide. The new system would be *engineered* so that it would be impossible to have head-on collisions and other types of accidents. Once a vehicle is tethered and programmed to be on a course no other vehicles would be in that lane.

I'm shaking just considering how to implement a project this huge and great. This idea came to me while riding an LA metro bus. Now is the time to think out-of-the-box—a totally new transportation system for the people. I believe our technology and engineering talents have evolved beyond what any of us could ever dream about.

If our ancestors could envision and build a national highway system, then it's our turn to correct the problems we know about today. I further believe the world is capable of building a worldwide tethered transportation system for cars, trucks, and buses without batteries.

The technology is available, skills are available, and tradespeople are available, including engineers who can handle the jobs for the task. Eisenhower started the U.S. highway road projects, and many people thought it was too ambitious. Some were concerned about the debt. EA were concerned about the environmental damage. Carbon wasn't even an issue back then. The EAs were not knowledgeable enough to forecast the carbon problem an oil transportation system would create.

Now back to the research. It's our turn to redesign transportation with nuclear for future generations. We all are certainly happy our previous generations moved forward with big governmental projects. The debt situation today reflects some of these projects.

The United States is the logical country to lead these efforts; we are the largest end-users of liquid fuels for transportation. Otherwise, the monarchies and oligarchies of the world can own (purchase) our brains, and/or Eastern countries will end up moving further ahead of Western countries, and they will have unlimited 2C power and no need for FF to limit growth or to purchase very expensive FF, all with 2C environmental temperatures.

It's important not to run out of FF before installing our world plan. After reading Bryce's new book on energy, I noted his conclusion "that we will not run out of FF because the first law of economics regarding supply and demand will come into balance." Of course, this is correct. Regarding the projections herein and those of the DOE and Shell are based on these three caveats:

1. Projected energy consumption needs are based on a 2% per year growth
2. Relatively stable energy prices
3. No major technological breakthroughs

Regarding no. 2, Mr. Bryce points out that prices will not be stable. What we bring to the table is what Mr. Bryce means by not having stable prices. If over the next 50 years we have the same increases in prices we had over the last 50 years, the price of energy will be:

Gasoline - $100.00 per gallon

Energy - $100.00 per million Btu equivalent to about $1.00 per kWh

And the price for nuclear will never be more than $1.00 per million Btu. And the generation cost of nuclear power will never be more than 10 cents/kWh. So the point here is that we and Mr. Bryce are correct. We will not physically deplete the last pound, but for the middlebrow of the world, billions of people will be without gasoline or cheap electricity.

Regarding no. 3 above, completing the technology for waste recycling and the IEAPR reactor are the new nuclear technologies that will void the caveat above, creating an environment to help Mr. Bryce's conclusion—never to run out of FF—come true.

Electricity: Even at today's prices, two billion people do not have access to electricity or cars. Does it matter whether or not there's still a few pounds left in the ground (and Bryce is correct)? No answer needed.

11.0

Final Sequitur

The fifth dimension to the matter was discovered when I was researching the economics of giving $7 billion to big oil industries for the food-to-fuel program and others. The science of economics and the work of younger economists, such as Mr. Piketty and his work using "big data," has solved the dichotomy that existed, etc. So this new knowledge regarding economics joins up with the quadrivium of energy, pollution, media, and scientists. He has proven via big data that untrammeled capitalism leads to the extreme income inequalities and other problems. And these problems will only get worse unless the public becomes knowledgeable enough to vote in politicians that will effect change. Another reason for truth in media, politicians, and villains, as Mr. Piketty would say. Otherwise, our freedoms and personal property rights are on the line. Read on for more information on the matter. This will lead you to the five Es of good governance.

12. One more (most important)

Sponsor a subgroup within the United Nations called the Energy Sustainability Council. One requirement will be for the IESC to issue an annual report on the state of energy and the progress made toward energy sustainability during the year. All member countries must agree that the report will be made available to the public.

13. Once the world plan (Chapter 5.3) is built, minimum FF will be needed. Transportation and power will be provided by nuclear and other renewables, providing electricity and a clean environment with 2C economics and environmental temperatures.

5.3

Summary of World Plan*

1. Execute the recommendations in Chapter 5.2.
2. Expedite energy conservation measures.
3. Discontinue all funding for biofuels (i.e., stop burning our food supplies). Our ancestors only burned shit for fuel. And China does today. Stop the fuelishness.
4. Expand solar and wind power, where appropriate, with minimum government funding; stop the fuelishness.
5. Expand nuclear power with at least ten new NPP in the next ten years for each large country (others prorated).
6. Develop infrastructure for electric transportation (including rail and buses).
7. Restart funding for nuclear waste recycling and other nuclear resource research. (IFR program)
8. Plan energy needs and resources for a minimum of fifty years.

*Each large country should execute this plan. Smaller countries should scale back accordingly (e.g., a blueprint for the world Renaissance).

9. Develop a score card for the voting public that rates a candidate's qualifications for office, based on the five Es needed for good governance.
   1. Energy
   2. Education
   3. Environment
   4. Economics

These first four were easy to name off, but what about no. 5?

   5. Equality

10. Continue supporting smaller nuclear plant development, especially modular units.

6.0

Jobs

This is a separate chapter written after an inspiration.

I was riding the bus this morning, 7:00 a.m., November 20, 2013, and sitting by a black man. He was telling me how hard life is without a job. He gets by through collecting bottles, cans, and any other junk worth anything. He uses the bus as his truck. He rides the bus a lot because, otherwise, he has to go back to his home under the expressway. It reminded me of my first paying self-employed job. I lived in Tabscott, Virginia, working the worn-out farm for my abusive father. Back to the job, I walked the roads around the farm and up to Route 250 where a gas station paid two cents a bottle. Cans were not used at that time. Anyway, our conversation inspired me to think of how this energy renaissance could create a job renaissance as well. We can develop a job scenario that is impelling. And this homeless man will get no credit or compensation. I'm crying as I think about this.

Let's start this journey together. First, my plan calls for building 10 new nuclear plants during the next 10 years but that's only a start. These 10 plants would service the growth needed for these years. Then we can start a reparation program to replace FF plants with NPP. Back to jobs: Each 1,000 megawatt plant cost about $8 billion, and a lot of the money goes toward good-paying construction jobs. Now, if we build twenty new plants during the next 10 years, then $160 billion will go into the economy. This would be about the size of China and India's future nuclear plans.

Again, these new plants would provide 2C energy. In our plan, we will be expanding our transportation system with electric cars, buses, and trucks and build all the new electric fueling stations needed. How many jobs will be created? Somebody else can get to this. The exact number doesn't matter; we know it's many. And we are not even counting jobs for the new transportation system if it's adopted.

So far, with the plants and fueling stations, we have added a couple of million jobs, at least. Then on to the railroads. Railroads are already the most efficient mode of transportation we have. This will require new jobs to build the electric grid system for the electric engines, as well as new jobs for manufacturing GE engines, etc.

All of these jobs will build out a 2C power network replacing the 4c FF network. This will save customers, maybe one-third of their electric bill and 80% of their gasoline bill. And all the new electric vehicles (for Ford and GM) will add more jobs. The 80% savings are without the new transportation system. With the new system, the savings are 99.44%.

And the workers of the world will be credited with saving the world. The next set of jobs will come from new processing plants to clean up the mess we have already made (e.g., scrubbing the CO_2 from the atmosphere with greenhouses and hydropond nurseries). This is a natural way to solve the imbalance, and it will supply the world with all the food needed to feed the starving children of the world. Can you imagine taking a substance that is killing people and our universe, and turning it into a fuel for our plants (the ones still alive)? As you can see, I got a little carried away (just a little). Now, we have to look at a couple of negatives.

There will be job losses in the FF industries (especially in the Appalachia) and among drillers. The rest of the world needs U.S. coal, so maybe it will not be so bad. Besides, we can increase our trade imbalance and use this money to fully compensate our workers. And I'm sure the workers would like to get paid without having to go into that hell hole. We just have to make sure we don't cause them to become lazy.

Again, I apologize for showing my politics. Besides, Obama will get his for stopping the funding on nuclear research programs.

Companies, like CNX and ACI, are already selling large quantities of coal to Asia and Europe, especially coal for steel. In the long run, maybe our plan will save the coal for future steel needs, a higher value use than burning the coal for power, especially since we have a 2C alternative to coal's 4c power without emissions.

An article in the WSJ (3/20/14) pointed to the direction Appalachian coal can go, to help replace coal that's currently going to power plants. Consol Energy Inc. (CNX), Arch Coal (ACI), and others have exported 10 million tons of coal to steelmakers around the world. Some of these plants across the world are:

Company	Country	Comment
Posco	South Korea	World's fifth largest
ArcelorMittal	Brazil	World's largest
Nippon Steel and Sumitomo Metal	Japan	Japan's largest.

So this answers some of the problems with job losses in Appalachia. This is what we come to when one has knowledge about the subject. Knowledge is so wonderful and important for the future. So the real issue is the supply situation for FF If coal is needed for a longer sustainable life for steel.

As pointed out in an NYT, 6/8/14, article, Appalachian jobs have been decreasing since 1980 due to:

1. Competition with gas
2. Cheaper coal from Wyoming that can be harvested with bulldozers instead of pick and shovel
3. Environmental regulations that were installed in the 1980s

Kentucky and other Appalachian states have already seen the light and are preparing for life minimum coal miner jobs, such as the Research Triangle Complex outside Raleigh, North Carolina.

What about the EA and EV?

Ralph Nader and other EV will lose Big Oil jobs. I'm sure they will fight on. And Big Oil has a golden parachute policy for employees. Since you are still on this train (except Ralph), let's consider the oeuvre of the whole EV movement against nuclear energy (e.g., the effort funded by Big Oil). Here, the EVs are against 2C energy that has killed only a handful of people while 4c energy has killed millions and is rapidly destroying the atmosphere. These EV must stay on this path because they will not have sponsors without Big Oil money. Prior to the start of this research project I was a Jane Fonda antinuke being duped by the propaganda in the 1960s. It was only after I had a stroke that I had time for this research project.

The biggest obstacle to all this boils down to one issue: "safety of nuclear plants." The CNN special *Pandora's Promise* shown on 11/8/14 and our findings herein should dispel your fears and point the way to cheap 2C energy. This passage is in Chapter 19 regarding the *New York Times* article on 11/5/13 by David Ropeik. David gives us a summary look via his book, *How Risky Is It Really?: Why Our Fears Don't Always Match Reality."*

As you can see, I could go on and on about this 2C energy renaissance, but first I will apologize for using terms that you aren't familiar with, like 2C and 4c. I guess you will just have to read on.

Another big concern is the waste (spent fuels). That's why we must continue funding the IFR and other modular small plants (more jobs). With this funding, our scientist and engineers will lead the way for the recycling of the spent fuels again and again until there's no more shit (more jobs).

And if you think that the engineers and scientists of the world can't accomplish this task, just think of Neal Armstrong walking on the moon. Which task is greater? Don't underestimate the brains our Lord through Jesus Christ has given us. We didn't come up with this knowledge on our own. Now God is telling us "it's your time to act before you destroy my creation."

Bye, bye, devil! We know the risk path we want to be on.

One final thought, I promise there will be so much money left on the table that we can scrape it up and use it to bail out Obamacare and increase the food stamp program and other health needs of the world. As many have stated, "no energy or expensive energy is the fastest pathway to poverty." Let's change paths. Today.

What we need is another John Kennedy and a Congress that believes in the greatness of the United States via a big collective government. (Sorry, I couldn't help myself.) Other big countries already know the power of big progressive governments. (I apologize again).

I hope this work is like Lincoln's Gettysburg Address, short but on target!

Now we have six reasons to get started today, with time is of the essence for:
1. Jobs
2. Environment (2C)
3. Extension of FF for 25,000 years
4. Economy

5. Feed the world with all the food we saved from not burning it up
6. Cheap 2C energy

Just like Martin Luther King, a little solitude in jail can focus your thoughts and create the reasons why

"We Cannot Wait"

Gee, I just stole this from the recent immigration movements of today.

Now that this is finally coming to an end, you can see how passionate I am about my unemployed brethren. If this book flops, I will not only be unemployed but broke, living on the streets of Watts, California. It's a sad story. And Big Oil says, "You deserve it for wasting your retirement money on such a project."

A short story: I got off the bus at the corner of Century and Central. I had a new bag (not a woman) with me and was headed toward a group of unemployed, young black men. Two Latino men I passed while I was walking asked me if I knew where the hell I was. I said, "Well, generally, and I have my walking stick with me." And he said "You are in the middle of 'Watts,'" code for "you better get your white ass out of here." Another angel in my life saved me. I understand street language. I caught the next bus! I didn't even care where it was going.

7.0

Introduction

Our politicians, monarchs, and oligarchs talk about growing our economy and those of the world, but few discuss how to power the growth and how to plan so that future generations can live in a world with diminishing or very expensive FF. With the FF industries and big oil managing our resources, what else do we expect except expensive FF. We will consume the last expensive pound of FF without planning for the future, because the industry knows that the FF prices will be very expensive as the supplies are exhausted.

We got into this situation because our Lord created the evolution process so that we could become knowledgeable enough to invent machines (and toys) so that we don't have to work (physically) so hard. I'm convinced we can work a plan, once we recognize the problem and work on the world plan beginning now.

So who am I, and how did I get here? I'm a chemical/petroleum engineer/project manager from Auburn University with a master of arts degree from Xavier. I have worked for 60 years, becoming a partner and vice president at the age of thirty-eight and the president of the Ohio Association of Consulting Engineers. I'm a licensed professional engineer (PE) in five states. I wrote a book for the DOE titled, *The Technology & Economics of Motor Fuel Ethanol*. I have designs for converting coal to liquid fuels (UCC project). I was raised on a worn-out farm in Virginia (Appalachia). We did not have electricity or running water. We had wood (not even coal) as fuel for cooking and heating. We had a well for water (hand-drawn). I lived a hard life as a child, without these fuels. It's only because of the grace of my Jesus that I am here to write this plan.

First, I was directed to study chemical/petroleum engineering as a co-op student at Auburn. I paid 100% of my education and graduated magna cum laude (top 3 in my class). Then I started my work career with Procter & Gamble, finishing as a VP of technology with Belcan Engineering (Cincinnati, Ohio). I suffered a stroke in the prime of my life at the age of sixty-two. I have survived, and now I have the time, health, and direction to put my experience and education together and write this book. At the same time, I am considering how to enter politics to help guide this plan. Another point is this: The stroke has helped me to intensely focus on the matter until I get to the matter's essence. I have worked as an energy engineer and large-project manager throughout my career.

The pending energy crisis will affect everyone. In fact, the rich may be affected more than the poor. The poor are accustomed to physical work, walking, staying warm with blankets, and sweating in the hot summers, unless this plan is implemented or a simulacrum occurs. The difference is that the rich can afford $100/million Btu.

When this book is finished, I will be sending a copy each to:

	Name	Who they are
1.	Mr. Obama	President (still)
	Mr. Moniz	Energy secretary (I want his job.)
2.	Bill Gates	Owner, TerraPower
3.	My brother, Charlie	DOE program manager,
4.	My niece, Kim Burns	The first PhD nuclear engineer from Georgia Tech. The dean told her when she was being interviewed: "It's too bad you are not a boy."
5.	Warren Buffet	Owns a lot of railroad businesses
6.	Harold Schulz	Starbucks.
7.	My son, John	The conservationist
8.	Mr. Macfarland	Assistant secretary, nuclear energy chairman
9.	Rachel Maddow	MSNBC
10.	Mr. Stone	CNN
11.	Ralph Nader (just kidding)	EV
12.	Al Gore	Nobel Prize winner and should have been President
13.	Paul Krugman	Nobel Prize economist, was the best before Piketty
14.	J.J. Suarez	Previous boss, Puerto Rico
15.	??	I will send a free copy to your representative.

16.   Magic Johnson   Starbucks, south-central Los Angeles

I'm not writing this book expecting returns that would pay for my time but just *hoping* to recover my expenses. This is the best I can do:

1. for future generations, for whom it would be a total disaster not to have cheap energy to power the world economies and worst, the two billion people without any electricity today
2. to stop the destruction of our atmosphere and planet
3. to plan for a world to have cheap 2C energy for at least 25,000 years

8.0

Bio-Background Continued

I was elected president of the Ohio Association of Consulting Engineers in 1972 at the age of thirty-eight. My other accomplishments are:

1. A pilot with 2,500 flying hours and 1,500 hours in instrument conditions; another reason why I'm lucky to be alive to write this book.
2. In 1990, I got my master's plumbing license and owned a company (Mr. Rooter franchise) in 1987 at age forty-three.
3. While at Auburn University, I achieved these honors:
    a) Elected to Tau Beta Pi, engineering honorary
    b) Elected to Phi Beta Kappa (top academic graduates)
    c) Elected to Phi Lambda Upsilon - chemistry honorary and was president in 1966
4. One of the top three graduates in 1968
5. I hold three patents (two in energy).

This is a portion of what I have achieved. If one lives long enough, many accolades can accumulate. This all adds to the power of knowledge and experience. A death means the death of power (Chapter 21). I have had so many friends whom have passed on too early. I want to make use of every minute my God through Jesus Christ has made available for me. This book is just the middle of my life, and if we all work together on this project, we can leave a legacy for future generations, just like what Rachel Maddox says on MSNBC: "The United States of America can still do great projects, like all the great accomplishments of past generations." And we say, "The world can do huge projects working together."

When pundits say we must solve our debt problems for our children and grandchildren, we say, "We are smart enough to solve the debt problem. And at the same time, take on great projects for future generations." Look around and see the great projects our ancestors left for us. What great projects have we completed in the last 20 years? Not much, except the Watts nuclear power plant. So let's all unite and solve the energy and environmental problems for future generations. Who cares what the debt is? With these problems solved, any debt is not big enough and can be repaid with energy and health savings in less than 20 years.

I ended up in Los Angeles, California, next to Watts, living out of my Ford 150 Econovan. When I had my stroke, I lost my job and my real estate in Dayton, Ohio. I was afraid my wife would kill me. A side effect of a stroke is extreme insecurity. I left our condo in Beverly Hills so that my daughter could keep her home. I moved to Ohio. My second ex-wife in Ohio helped me get set up. Then I purchased a condo in Cincinnati for my daughter and me. My daughter had no interest in moving to Ohio. She and I had visited Ohio several times and she said it's too cold and she would miss all her friends. I agreed. My wife promptly filed divorce papers. The divorce was final in 2009 (joint custody). I could not live in Ohio without my daughter. I moved back to LA in 2010. My daughter and I are together at least twice a week. Her Ima is a very good mother. So that's the story. I was close to suicide a couple of times, then my Lord whispered to me and told me, "I have a whole pie full of goodies for you."

This project (book) is just one slice of it. I can't wait to see what the other slices will be.

9.0

World Supplies of Fossil Fuels (FF) Resources

The best news is that the United States has the most resources in the world with 5,619 quadrillion Btu (known). This represents 17% of world resources. The worst news are:

1. The United States leads the world in energy consumption, especially oil (see Chapter 10).
2. Petroleum resources (oil) are only 5% of the total.
3. The world consumption of FF resources is increasing rapidly.
4. The United States is expanding the sale of coal and natural gas (NG) to other countries.
5. The United States consumes more oil than the total of all the oil consumed by the other top three countries.

So with these negatives, world supplies of FF are diminishing fast. The only inexpensive way to stop the rapid consumption of FF is to build the nuclear program recommended herein.

Very expensive FF is another way to slow down consumption. Of course, other renewables could do the same thing (see Chapter 11), except these renewables do not meet the criteria regarding inexpensive and available supplies.

Table 9.1 shows the FF resources of the world. This information was extracted from several DOE reports via:

1. Congressional research resources (CRS) reports
2. Energy Information Agency (EIA)
3. An update on FF supplies as of 2014 is included in Chapter 17

9.1

World supplies of FF (Ej Btu)

	Oil	Coal	Gas	Undiscovered oil/gas	Total Ej
USA	124	5,251	244	2,039	5,619+
Russia	?	?	?	? 1,900	5,300+?
Canada	1,030	146+	60	1,242	?
Mexico	?	?	?	?	?
China	93	2,522	82	2,862	3,900
India/N 2				1,242	1,242
Iran	803	30	975	663	2,000
Qatar	931	NE	NE	?	?
Iraq	500	NE	NE	397	1,179?
Saudi Arabia	1,500	NE	Unknown	?	1,500?
Australia	?	2,946	?	?	?
Venezuela	1,000	NE	?	?	1,000?
UAE	1,000	NE	?	?	1,000?
Nigeria	900	NE	?	?	900?
Brazil	200		200	300	700
Former USSR					1,200
Others	500	?	?	?	?
Total Ej					37,880?

(2) – includes New Zealand and India

Fossil fuels represent the carbon resources from the carbon ferrous period, going back 300 million years, basically representing *all* the solar energy stored during these years.

The table information is primarily from the EIA (Energy Information Administration) and Congressional research resources (CRS).

Crude Oil + CH3 liquids
World Energy Council 2004

                Barrels (Equivalent) Ej ×100

Saudi Arabia	262.8
Iraq	115.0
Iran	99.1
UAE	97.8
Kuwait	96.5
Venezuela	77.3
Russia	60.0
Libya	36.0
Nigeria	31.5
USA	30.7
Rest of world	192.0

6 mm Btu/barrel of oil. Btu = 6.1×10^9 j
26 mm Btu/ton coal – 2/85% efficiency = 3.2×10^{10} j
Ton of uranium = 7.4×10^{16} j
1 BD e = 5.63 Gj
1 ton e coal = 25 Gj

U.S. production
- 1965 – 9.0 mm barrels/e/yr
- 2005 – 15 mm barrels/e/yr

U.S. consumption
- 1965 – 12 mm /e/yr
- 2005 – 20 mm/e/yr

Jan. 1, 1996   DOE fossil fuel resources
    Total – 37,880 Ej
    CH3 – 14%
    Oil – 17%
    Coal – 69%

10.0

World Energy Consumption and Projections*

World Energy Consumption Per Year (Btu)

Total	Oil	Coal	CH3
1970 - 240 Ej	120	70	50
1995 - 330 Ej	140	80	80

*EIA (USA)

The supply will be exhausted in 2074 with a 2% growth per year, according to DOE in 2005. Again, Bryce refutes these projections because energy will get to be very expensive, stopping growth and the 2% annual increase. Even if Bryce is correct, we will still have an *oil* crisis within 30 years.

With these projections and supplies, the exact numbers are not the issue. No matter what, even if the numbers are off by 100%, we will run out of FF by 2095 instead of the 2074 projection. So our grandchildren will be young when this happens. The world discovered FF at the beginning of the Industrial Revolution, about 150 years ago. In these 150 years, we have consumed about 50% of the known FF supplies. We are on this safe path to consume all the FF in 100 years. The world is capable of arresting this by embarking on a plan to manage our energy supplies for any time period desirable. Our plan shows how to do this for 25,000 years. After all, humans have been on this planet burning fuels for about 10,000 years. So the 25,000-year plan is reasonable for today. After completing the nuclear research outlined in the world plan, world energy supplies can be extended for another 250,000 years. Wake up, you are not dreaming.

How much energy we are consuming today and projections for the future

World capacity: 367 GWe (now)

Expanding to 524–740 GWe (2030)

Scenarios*

2005	2010	2015	2020	2025	2030	Year
367	372-389	372-447	367-518	371-613	281-740	GWe

* From EIA

World energy consumption: 474×10^{18}j with 80-90% from FF nuclear 6% and Biomass at 4%.

Sun energy can generate 1366 watts/m²

So a solar power plant like the one in CA takes 127,4000,000/km² 1740×10^{17}/km² for a (petawatt) plant. With 500 ac. of panels.

*I did the calculation without a calculator, so I'm sure someone will send me the corrected value in acres. Total plant area: 3,500 acres (5 sq. mi.)

China is the largest FF consumer of electricity at 100 units Btu versus the United States at 80. The United States consumes more oil for transportation than China by 135 to 20 units Btu.

And China's nuclear power is growing at 10% per year versus the United States's 1%. China has plans to build 10 nuclear plants per 10 years.

Source: NYT, 10/27/13

World Electricity Generation Change (yoy)

Coal	1.4 billion megawatts -	13.6%
Gas	1.1 -	23.6%
Nuclear	0.7 -	2.5%

Source: Energy Department, reported in the *Wall Street Journal*, February 11, 2013

10.

Energy Consumption and Projections (Shell Oil)

Again, the base line is from DOE information and a realistic scenario of the world's energy consumption up to 2050 as projected by the Shell Oil Co. (see www.shell.com/scenarios.com).

Table 10-1 Shell Scenario Ej*

	2000	2010	2020	2030	2040	2050
Oil	147	176	186	179	160	141
Gas	88	110	133	134	124	108
Coal	97	144	199	210	246	263
Nuclear	28	31	34	36	38	43
Biomass	44	48	59	92	106	131
Solar	0	0	2	26	62	94
Wind	0	2	9	18	27	36
Other renewables	13	19	28	38	51	65
Total primary energy	417	531	650	734	815	880

** Shell has a conflict of interest when projecting nuclear's future. The oil industry doesn't support a nuclear expansion because nuclear would displace FF. Nuclear power is cheaper by 50%. Look at what happened to the nuclear plant built on Long Island and never opened because the FF industries teamed up with the EV and spread lies/fear about nuclear. See CNN's special *Pandora's Promise* and Chapters 12, 19, 20, and 23.

It's interesting that Shell and the government agree on projections for future energy needs. But the U.S. Department of Energy doesn't have a plan on how to get there. Our projection for nuclear plants should be more like 300 in 2050. This doesn't include any conversions from FF to nuclear. That would be in addition to the 300, with the world plan herein. If this were to come about, nuclear power for the United States would be ~30% up from 17% today. Some would say there's no way we could add that much nuclear capacity in just 40 years. We say, "If France can expand nuclear power from 10% to 80% in 20 years, the United States can accomplish the 30% expansion in 40 years." Being an engineer and project manager, I say, "This is a piece of cake."

Even with Shell's optimistic projections, the only fuel available to fill the gap is nuclear, and it will have to increase from 28 Ej to 300 by 2020. This will take 10 new N.P.P. for each large country in every 10 years; just to keep up with demand. FF companies would never support or believe nuclear could support the growth.

11.

Power Generation Cost (All Fuels)

The Shell scenario, Chapter 10, shows:
1. Currently FF represents about 75% of U.S. fuels consumption
2. By 2050, it could be reduced to 60% (Shell scenario)

Renewables* currently represent 8% of all fuels, with 7.744 quadrillion Btu per year from a fuel total of 94.578 quadrillion Btu per year

The 8% renewables are:
1% solar
20% biofuels
9% wind
5% geothermal
6% biowaste
24% wood
35% Hydropower
*From EIA (Energy Information Administration)

Nuclear fuels are 9%

The Shell scenario, Chapter 10, shows
1. Renewable growing from 8% to 20% in 2050
2. Nuclear growing from 9% to 15% in 2050

11.1

Food to Fuel

Shell discusses the many problems with increasing biomass energy from 1.6% to 3.2%. A recent *New York Times* article (4/7/2011) discusses how the rush to use crops as fuel raises food prices and hunger fears throughout the world. Some highlights of the NYT article:

1. Cassava root, 98% going mostly to fuels
2. Corn prices increased from $2.60/bu. to $8.0/bu. last year because of biofuels
3. Sugar - Brazil and South America
4. Rapeseed used in Europe
5. Palm oil in other parts of the world

All of these food-to-fuel programs require huge taxpayer support. Some countries don't have any choice, but here in the United States, we have a choice. But we have chosen the path that cost taxpayers roughly $27 billion dollars per year only because the oligarchs have convinced Congress that this program saves us from purchasing foreign oil. But at the same time, they hide the fact that this enriches themselves with the $7 billion dollars of direct tax subsidies.

Gasohol (alcohol fuel) update: 01/20/14 (research)

I decided to do some more work on just how much money is given directly to the oil companies that blend the alcohol with gasoline. The total is ~$7 billion per year. I couldn't believe the numbers. Here are the facts that support the numbers.
1. In 2012, we purchased 13.3 billion gallons of gasoline
2. In 2012, we manufactured or purchased 13.3 billion gallons of industrial grade alcohol.
3. 70% of the alcohol goes to blend gasohol.

4. The federal and state taxes that are collected and given to the oil industry are:
   a. federal - 18.04 c./gal.
   b. state - 23.5 c./gal. (average)

And of course, the agri industries like the program because of the higher food prices. So the public gets hit with both barrels: taxes and higher food prices. President Reagan was the only president that didn't support the program. But even he wasn't powerful enough to fight Big Oil and the agri industries. To be fair, the money actually ends up going to the agri industry and processors of industrial alcohol, not to Big Oil.

After writing our findings on the direct tax subsidies given to the companies that blend the alcohol fuels, Bryce wrote in the NYT, 03/10/15, to "end the ethanol rip-off." He based his comments on the energy differences between ethanol at 76,000 Btu per gallon compared to gasoline at 114,000. He got into the numbers and showed this cost consumers $10 billion per year. He doesn't try to determine the cost to consumers for the increase in food prices due to the food-to-fuel program. This program was initiated by Congress in the 1970s. What's astounding to us is this: Bryce doesn't say *anything* about the direct tax subsidies we described in detail above. We can't think of any reason why he didn't discuss the direct tax subsidies other than some kind of conflict-of-interest. Bringing these direct tax subsidies to the public's interest is risky because so many wealthy power groups have a stake in this program continuing.

But we are here to add up all the numbers.
1. Direct tax subsidies of $7 billion/yr
2. Indirect cost regarding energy differences of $10 billion/yr
3. Indirect cost due to higher food cost. Since the latter can't be easily determined, an opinion we have is that this latter is at least equal to the $10. billion above.

So with A + B + C, the total is $27 billion per year. I totaled these for all the anticlimate deniers since they can't add 2 plus 2 equals 4 or answer the question: Where did all the CO_2 come from if not from human activities?

To continue, since this has been ongoing for almost 50 years, the cost to the U.S. public for this one policy mistake is ~$1.35 trillion. And as of this date, 3/10/15 and after breaking the highway trust fund, no one in Washington is talking about eliminating this. Such a wasteful, expensive program.

Many countries are mandating biofuels be increased. The unintended consequences of much higher food prices will more than likely cause politicians to correct these mandates in the future after the public suffers from higher food prices.

This year's (2015) Wednesday is Earth Day. One issue that is big this year is food wastage. According to Dana Gunders, an activist for Earth Day, we are wasting about 40% of agriculture crops because of losses and waste at home and in industries (NYT 3-12-15). One waste she missed is the 40% of corn crops lost to the food to fuel program. Maybe the drought's we are suffering in California, Brazil, and others will wake us up and *stop this fuelishness*. So what options do we have? Other renewables or nuclear? You will have the answer soon.
The Shell scenario suggests:

1. Solar from 0% today to 94 Ej in 1950. Our opinion is that there is no way solar will increase this much for two primary reasons:
   a. It's too expensive (more than 250% more than nuclear.
   b. It takes a backup fuel when the sun is not shining.
2. Wind from 0% today to 36 Ej in 1950 (see Germany, Chapter 14). Same reasons as solar apply here. Wind power is too expensive, and areas to harvest are limited either from the sun or wind for the energy needs that will be needed over the next fifty generations.

11.2

Solar

The Natural Resources Defense Council (NRDC) reported that electricity from small to medium installations costs around 12–30 cents/kWh, and have fallen sharply in the last 20 years, and this trend should continue in the future.

Tax subsidies, and direct government subsidies allow for these solar installations to be built. A large solar power plant is going forward in California as the BLTHE plant. This is the largest solar plant to come on line in 2013. According to Sanda National Labs, the generation cost are "expected" to fall to about 5 cents/kWh by 2020 from 8cents today. This solar power plant has all the best features available for solar energy:

1. Sun shining at a direct angle for more than twelve hours per day and being on the edge of the Mojave Desert in California and Arizona
2. Cheap land
3. Inexpensive labor
4. Relatively easy installation of transmission lines close to populated areas
5. The 5 cents/kWh by 2020 assumes the best results for all variables that are unknown at this time. So this estimate is the very best price for this plant with the best of all variables. And the cost for power is still 250% costlier than nuclear.

And now, with the last entry, 12-4-15, in both the NYT 7 WSJ, Mr Crane, the CEO of NRG is stepping down after 14 years. In the last year NRG stock has tumbled from $30.00 to $10.00. The reason: to heavy investments in renewables. This is even with large government investments in solar and Wind energy. Mr Crane should have procured one of his engineers to put pen to paper before making the large investments. As we say "its not the technology that needs to be proven; its the economies." The United States government has invested some $1.6 billion in NRG solar projects.

Solar Energy Update - 2/14/14, NYT

The analysis for solar energy to provide 5c power in the future was based on the solar energy plant in Nipton, California, called the Ivanpah Solar Power Plant with five square miles of solar panels (3,750 acres); panel surface face area is the 100 acres. Taxpayers put up $1.6 billion of the investment, and the power plant owners, NRG Energy, put up only $300 million. Solyndra cost taxpayers another $400 million. The energy secretary, "Mr. Moniz," has stated that "the government has funded this plant so that the technology can be used by private companies." NRG and Mr. Moniz have speculated in the past that the technology will provide 5c power once the plant is running for a few years, after the government has spent over $2 billion dollars of taxpayer money. NRG has been asked by Bright Source to go in with them to build another solar plant. NRG refused, stating that the technology is not proven yet to be competitive. Also Mr. Moniz stated that the only hope for this technology to be competitive is by constructing large plants, not small plants.

This proves our analysis regarding small plants. This solar plant is large enough to supply electricity for 140,000 homes, as a 1.4 mega kW plant could. Bright Source Energy would go in partnership with Bechtel the large engineering and construction company located in Houston, Texas. A big problem NRG has is that, NRG has to provide FF electricity (not wind) when the sun is not shining. So if NRG has to have all the infrastructure in place to provide the 4c FF electricity, why would NRG have the incentive to fund a solar plant when the best fuel is nuclear? I'm ahead of myself here—just keep reading. NRG already operates several 2c nuclear plants. The only reason NRG would not build a new nuclear plant is because permitting is so difficult. EV, et al are walking in front of bulldozers.

The solar power plant in California is already proving that solar energy is only viable if taxpayers fund the plant. No one in the DOE (Mr. Moniz) has the courage to step forward and tell the president that not only did they waste $500 million on Solyndra, but now it looks like they have wasted $2 billion on Ivanpah. And now this leads us full circle back to 2C electricity that's already proven and ready to go. The only barrier is the fear and lies of the EV. Come on, world, let's not let fear and lies control our destiny. Again, our Lord has provided us with the brains to go forward. It's up to us to use what he has provided.

And now Japan's problems with solar energy have made headlines on the first page of the NYT on March 4, 2015: "Short-Circuiting a Solar Boom in Japan." The utility industry in Japan is balking because of rising solar cost. What's actually happening, even though the article doesn't specifically says so, is that the Japanese government is cutting back on subsidies for solar, expecting the utility industry to carry the burden. Companies like Kyushu Electric Power Co. announced that it would stop purchasing electricity from the solar ventures. If all the solar facilities on Japan's docket today were built, *users* would have to pay ¥2.7 trillion a year in special charges. Eventually, the increased cost of expensive electricity rises to the surface. Government subsidies can only hide the cost so long. The good news from this is that "this is forcing Japan to reassess its decision to close all the nuclear power plants."

11.3

Wind Power

The generation cost for wind energy is difficult to determine because it depends on location, size, and other environmental issues. For example, just like the German situation, after spending $50 billion for the windmills and generators, the country now must spend $27 billion on transmission lines.

A better comparison is the Cape Cod wind projects and the marketplace in Massachusetts, which just signed contracts to purchase land-based wind power from Maine and New Hampshire for 8 cents/kWh and wind over the ocean from (Cape Cod), which is being contracted at 19 cents/kWh. And even at these prices, the electricity delivered is uneven because the wind is uneven (obvious). Other energy sources must be used to even out the electricity supply. (Source: NYT, 10/23/13)

In the United States, wind power in 2012 increased by 13 gigawatts with government subsidies; then in 2013, with minimum subsidies, only 1.0 gigawatt was added. The World Resources Institute (WRI), via Letha Tawney, stated that with "any uncertainty about the production tax credit (PTC), it all stops." The International Energy Agency (IEA) noted that the deployment of photovoltaic solar and wind-powered electricity was meeting goals set by the WRI, such as keeping the temperature rise below 3.6 degrees Fahrenheit (2 °C). Keeping the rise to this level will avoid a truly disruptive climatic upheaval. Other renewables such as bioenergy, geothermal, and offshore wind were not meeting goals (NYT, 11/19/14).

The U.S. IEA reported that with the latest technology, onshore wind energy can be supplied at a cost of only 7 cents/kWh as compared to 5 cents/kWh that was projected five years ago. Also, the EIA reported that the latest projection for photovoltaic solar electricity is only 8 cents/kWh. These new prices/cost are being reported because of the resetting of energy cost due to the agreements recently (11/10/14) reached by in the United States and China. With a carbon tax of $25/ton starting next year (2015), the United States still can't meet its agreements regarding the reduction of carbon emissions. This tax will cause electricity to cost 2 cents/kWh more. So the cost for FF electricity will be 6 cents/kWh. For renewables, this means the cost for FF is almost equivalent to that of the best renewable (solar), and biofuels have fallen off the chart.

Wind Subsidies in the United States (WSJ, 12/1/14)

For almost 30 years, the PTC has cost taxpayers over $70 billion in the last seven years and will cost taxpayer's over $2 billion dollars, each year going forward. The PTC is 2.3 cents/kWh, the same generation cost as nuclear power w/o subsidies. What kind of scared government do we have anyway?

More:

Total subsidies in 2010:
    Wind - 5.6 cents/kWh
    Gas - 0.6 cent/kWh
    Nuclear - 0.03 cent/kWh

Note: This does not include food or solar subsidies. All these subsidies go to already profitable companies. And the U.S. Congress is reducing food stamps and unemployment payments. What kind of fuelish government do we have anyway?

Bryce Article in WSJ, 6/12/14
Global hydrocarbon consumption, 218 million barrels/e/day
    83 - oil
    75 - coal
    60 - gas

He cites a Mr. McKibben, the world's most famous environmentalist. Mr. McKibben wants to decrease CO_2 levels to 350 ppm. To do this, the world would have to reduce FF consumption by twentyfold. With energy growing by 450 terawatt-hour per year and the International Energy Agency expecting this demand to continue for several decades, for renewables to satisfy this additional capacity:
1. Solar - The world would have to install sixteen times the installed capacity of Germany's 33,000 megawatts every year.
2. Wind - Again, to supply just the growth every year, the world would have to install 108,000 square miles of wind turbines. Today, the United States has about 25 square miles of wind-harvesting turbines.

Wind and Solar Power Investments in Germany

The German state of Schleswig-Holstein has constructed wind turbines and solar panels to produce 12,000 megawatts of power. This cost about $50 billion, and now another $27 billion must be spent to construct 1,700 miles of high capacity power lines to deliver the electricity to German consumers in the south (NYT, 9/19/13). And now more subsidies are needed to suppress the power cost.

And it's not just Germany that is wobbling on strict sanctions on Russia. Big Oil in the United States is applying the same pressure on Obama because companies, like Exxon Mobil, are heavily in Russian oil. These oil companies, like LUKoil, Rosneft, and Gazprom, are owned by friends of Putin. Russia and the United States have a lot in common when it comes to who has power within their respective governments. Russia is a monarchy with oil controlling the president. The United States is an oligarchy with oil money controlling our politicians. And the EV are their ambassadors (mercenaries). Why else would taxpayers be funding the food-to-fuel program that cause food prices to go up and up. It's just one example again of the EVs fostering the lies/fear.

The public guides our politicians based on EV fears and lies. Germany is the most productive and competitive country in Europe. Once Germany completes its conversion to renewable products, cars, etc. will be more expensive. Germany continues to raze existing villages so that the coal under certain towns can be mined. Afterwasch is such a village; it has 25,000 people. This is taking place because Germany needs the coal until the conversion to renewables is complete; all because of its decision to shut down the NPP.

Let's all be thankful the Obama administration is not quite as fuelish as Merkel's administration. But be aware, it's the public's attitude on energy that drives the decisions in both countries. The United States has had presidential candidates in the past that would have shuttered all of the country's NPP. So it's not a stretch to say that the United States could make the same mistakes.

Since Merkel declared that Germany would close all nuclear plants in the next few years: The chancellor made that decision after Fukushima and has spent billions to install renewable energy. The price of energy has skyrocketed, causing Germany to not only subsidize the investments in renewable energy plants but also now subsidize electricity prices for residents and businesses. If Germany stays on this path, I forecast it will lose its competitiveness. (9/1/13)

And just this month (10/13), Obama proposed a plan to spend another $8 billion to build renewable energy power replacements. Nuclear plants can be built without this much of a subsidy. And the power is cheaper. The last nuclear plant in the United States cost $5 billion, but it took 20 years to build.

Energiebilanzen, an association that tracks energy projects in Germany reports that Germany has converted 23% of its electricity needs from nuclear to renewables at a cost of over $100 billion and will spend up to $550 billion to reach 80%.

And again, Germany (NYT, 4/27/14):

European private companies and governments are resisting imposing UN (U.S.) sanctions against Russia for taking Crimea and for the Ukraine situation. Notably, Germany has a large trade deficit caused by its need to purchase oil from Russia to fuel the power plants that were recently converted from nuclear to FF. France is the only big European country that wants Europe to go along with the UN (U.S.) sanctions.

An example of how this is playing out: Mr. Steele, the CEO for the Wintershall division of the huge chemical company BASF, advised Europe to not turn away from Russia, and that Russia is already being punished enough with capital fleeing the country since Crimea. BASF is deeply involved with trading oil and gas for chemicals. The import–export trade situation for Europe is $370 billion in 2012.

Sanctions are our only way to keep Russia from taking over Ukraine. Does this mean that if Russia makes this move, we have another war over oil? After all, the United States has already gone to war with Iraq, Afghanistan, and is threatening Iran and Syria. Look at what the monarchs and oligarchs have done and are doing to the world. And which huge country is not involved with these wars and possible wars? By now, I think you know the answer to that question. If not, read Chapter 14.

All of this supports our conclusions that there is no way renewables can satisfy even the growth in energy, much less reduce FF consumption.

11.4

Fossil Fuels.*

Summary Generation Cost/* (background for cost herein) cents/kWh

	5% DCFRRI (cents)	10% DCFRRI (cents)
Coal	25-50	35-60
Gas	37-60	40-63
Oil**	Not competitive with gas or coal	

*Discounted cash flow rate of return on investment (DCFRRI)
*EIA (Energy Information Administration)
**The cost for generating electricity from oil is not calculated because it is not competitive with gas or coal FF

Investment cost for FF per megawatt (large plants)

Coal - $30,000
Gas - $15,000
Nuclear – $90,000

Energy generated from FF in 2012
Billion megawatts
Coal – 1.4 (14% increase from 2011)
Gas – 1.1 (24% increase from 2011)

Here's another story. Back in the good old days (Mac) before Apple and apps, if you had a computer, you would have to do your own programming. My friend, Kurt Kalpish, and I programmed the app for DCFRRI analysis in Fortran IV. Actually, Kurt was the better programmer, and I was the lazy one (No, Rommey, I wasn't on food stamps. I had a job.) The program was used on our projects for energy and the book, *Technology and Economics for Motor Fuel Alcohol*. One aspect of the program that was especially difficult was considering the method of depreciation. All utilities use the sinking fund method. This front-loads the depreciation for lower taxes.

*So with these two computations: the rates of 5 and 10 are conservative; as the interest rate goes down, the cost goes down, and 5% is a good rate in today's environment.

We sold a couple of these programs to utility companies.

Fossil Fuels and Renewables

Power Fuel Prices
    Price per million Btu
    Oil - $25.00
    Gas - 3.50
    Coal - $3.50
    Nuclear - 2 cents
    Wind - free
    Sun - free
    Fusion - 30 years (more research)

Resources (supplies) at 2% growth rate)

    Oil - 30 years
    Gas - 50 years (updated)
    Coal - 100 years
    Nuclear (present technology) - 25,000 years
    Nuclear (completed technologies) - almost limitless
    Wind - limited "areas available to capture"
    Sun - limited "areas available to capture"
    Fusion - limitless
    Power generation cost (large plants) - Table 11.7-1

As reported in the WSJ, 7/14/14/ in an article by Bryce on his recent book on energy, "he confirms our cost analysis for gas and nuclear at 3c and 2c, respectively." Bryce promotes N2N: Power generation with natural gas then nuclear. Again this is similar to our conclusion. And we provide the evidence: numbers and plan for the world's nuclear build-out.

11.5

Nuclear

Economics:
Generation cost: cents/kWh, calculated using DCRRI method
 5% discount rate (cents)   10% discount rate (cents)
 21-23   30-50

Other information are in Chapter 12, especially sustainable supplies.

Energy generated from nuclear in 2012 - 0.7 billion megawatts, with an increase of 2.5% from 2011.

The Fukushima knowledge is already showing up in the United States. Exelon, the owner of 17 of the 100 U.S. nuclear power plants, is retrofitting existing plants with new trucks and pumps, just in case a Fukushima event were to happen. This work is expensive, but these expenses are pennies (cents) compared to the dollar savings over an operating life of 30 years; delivering power at 2 cents/kWh. This is the work of engineers, scientists, and other workers. We become God's ambassadors via his evolution process. As Exelon explains: "What if we were wrong" regarding our assessment of the seismic conditions at the plant locations. In other words, this is extra insurance risk protection.

The chairwoman of the U.S. Nuclear Regulatory Commission recently stated: "It woke everybody up," adding, "Hey, you didn't even think about these issues happening together. Now redesign all plants with the possibility of an earthquake that could create a tsunami and revamp the emergency diesel generators in case the plant is left without electricity for extended times." (NYT, (3/10/14)

So the regulators are imposing this "requirement" on new as well as existing plants. This is the job of regulators and engineers. It takes it out of the hands of crony politicians and NPT owners who know nothing about these matters. And the EV are too busy collecting money from Big Oil.

11.6

Fusion

Fission is defined as the splitting of atoms into smaller matter, releasing the energy from the bonding, millions of times more energy than that released from burning FF.

Fusion has these advantages over nuclear uranium:
1. Supplies - four times more than uranium
2. Energy released is more
3. Easier waste recycling, 95% less waste
4. Has an ultrahigh melting point, making it less likely to melt down
5. Resources:
    f. Thorium
    g. Some isotopes of uranium.
    Thor - god of thunder in Norse mythology

Disadvantages

1. 30 more years of research are needed and 10 years of prototype development
2. Highly specialized reactors; already more than 50% there

Research into fusion reactors began in earnest in the early 1950s. And was added to the Manhattan project, and continues to this day. Every year, there's a saying that goes around.

"We only have 30 years of research left to go." There are two projects being funded today in the United States and other countries, namely:

1. The national ignition facility
2. The ITER projects (see below)

This research is being conducted on the fusion between deuterium and tritium. This reaction releases 14.1 MeV and is consistent with Einstein's $E = \Delta Mc^2$, where Δm is the reduction in mass after the collusion. The construction of facilities to house the complex is scheduled for completion in 2019. The experiments will began in 2020, with full fission completed by 2027. That's how close we are to releasing more energy than the world would ever need. This will be the culmination of 75 years of research already completed. Once this research is complete, then the world will have plenty of jobs to implement the new energy source.

The caveat for any of this to occur is for the world to replace FF with nuclear and some renewables, before we destroy the earth's atmosphere!

This is not an idle threat, it's real! France, Russia, China, and India have already signed up. Other countries better sign up soon or they will be left behind with only the hot CO_2 to breathe.

11.7

Summary

Table 11-7.1

Summary Generation Cost (Large Power Plants)*

Fuel	Cost cents/kWh*
Nuclear	2.0 (rounded)
Wind	8.0 over land
Wind	17.0 over oceans (rounded)
Coal	3.0
Gas	3.0
Oil**	**
Solar***	5.0-30 (rounded)

Note: Bryce, in his new energy book confirms the gas and nuclear numbers of 3 and 2, respectively. We got our cost numbers from the DCF ROI analysis provided by the DOE EIA.

*Generation cost
**Oil is not competitive with coal or gas (the other FF)
***The 5.0 cost is based on the projections from the California/Arizona plant. The most recent information is that the plant is having trouble with the 5.0 cost. Data in Chapters 11 and 12

*Large power plants +1,000 megawatts)

Oil is used today to temper or for emergency needs, not for new plants.

11.8

Hydropower

I didn't include hydropower in the above analysis on fuels because I believed HP was already developed in areas that had the opportunities. With little research, here are the findings:

In the USA:

Hoover Dam generates electricity and sells it commercially for
- 2 cents pkwh (our 2C).
- It has a capacity of 4000 thousand kw.

Niagara Falls Dam generates electricity and sells to customers for
- 7 cents pkwh.
- It has a capacity of 2000 thousand kw.

The difference in prices (rates) is because
- Hoover Dam is twice as large (economies of scale)
- Different countries control how to set the prices.

China, being a big country with lots of rivers, has the most installed HP plants in the world, with 50% of the worldwide dams, or approximately 65,000, and dams 50 feet high. These power plants generate 280,000 thousand kw. China had very few HP plants before Mao. Other leaders have followed his direction.

Note: A thousand thousand kilowatts is a billion watts. Hoover Dam and the adjacent nuclear plant yield a total of four billion watts. The three Ivanghan Solar plants have a total capacity of 392 thousand watts. Soon to be derated because of clouds.

The above discussion on HP brings up a point regarding the size of projects, and the thinking of our leaders needs to be bigger than all this talk about little projects and the burning of our food supplies. Our energy needs are bigger than just relying on small renewable projects like the solar plant in the Mojave Desert known as the Ivanpah plant, a.k.a. the 392 thousand kw plant. In addition to the smallness of the plant, it is having trouble meeting its objective to generate electricity at 5 cents pkwh.

China had to relocate 23 million of its citizens to other areas and have flooded thousands of square miles of farmland/vegetation and villages. The long term effect on China is unknown. She will just have to live out the time before measuring the positives/negatives of Mao's leadership.

China, unlike the United States, doesn't prepare environmental impact statements. The main point here (before we go on to more on EIT) is that providing power to the people does not come with a free lunch. It's so important for politicians, activists, and the media to be honest with the public for them to understand what's in their best interest and express it to their leadership. Otherwise, it's a total misuse of free speech." The public will be misguided and would be better off without it.

Regardless of the environmental issues, not providing cheap 2C energy would be a total leadership mistake. The only option for China other than HP would be to build 10 big NPP. Now this is an option that needs to be open and transparent. This isn't just a past-gone-choice situation.

Africa (Chapter 14.14) is faced with this choice today, and hence our motto "A risk path choice going forward is better than a safe path going backward." Let's hope Africa makes the right decision. Let's keep the oligarchs' free speech out of this area.

And now with this work on hydropower, we see that HP is the only renewable power that can compete with nuclear. It takes installations bigger than Hoover Dam before 2C power can occur.

12.0

How Do Nuclear Fuels Fit into This Picture?

When I started research on this book, I didn't realize the importance of nuclear fuel resources and supplies. Thus the title of this chapter. After my research, I'm convinced that nuclear power is the *only hope* we have for a sustainable, inexpensive energy for the future, and therefore, a future with 2C energy for many generations. As I explained earlier, my background is with fossil fuels. During the 1960s, I was a typical Jane Fonda antinuke. I remember seminars that promoted nuclear power, but they could not answer the question: What about the resources and supplies availability for many generations? My research on fuel resources and supplies are shown in Table 12.1. Our research shows that:

1. We have enough of the easily available resources to power our future needs for 25,000 years with recycle. Once our science and engineering is complete:
   a. Recycling of spent fuels (this also helps solve the spent fuel problem)
   b. Fast recycling technology (IFR program)
   c. Advanced thorium fusion technology. The Russians plus others are funding this.

Then we will have resources and supplies for "unlimited".

2. Consider that these limitations are only:
   a. Knowledge
   c. Science
   d. Engineering

These limitations are problems we are good at solving. If you need a complete explanation on the technologies, see the sources on the tables.

The above comments were written before the Japanese Fukushima disasters. The statements are still true, but it will make it especially hard to educate politicians and the public that nuclear is the only answer. One example is Germany, Chapter 14.

Nuclear Material

A material that is capable of fission/fusion, following the capture of a thermal electron;
examples are U238 to Pu239 and Th232 to 233u. (fission.)

Table 12.1

Sustainable Years of Nuclear Fuel Supply*

Potential energy production	Energy Potential energy production conventional tWh^18	Energy Potential energy production world nuclear electricity generation conventional (tWh^18)	Years Sustainable years at 1999 world nuclear electricity total conventional	Years Sustainable energy at 1999 world nuclear electricity generation (total)
Current fuel cycle	827,000	21,200,000	326	8,350**
only one recycle fuel cycle	930,000	23,900,000	366	9,410
Light water and fast reactor with recycle	1,240,000	31,800,000	488	12,500
fast reactor fuel cycle/recycle	+26,000,000	+630,000,000	10,000	25,000
Advanced thorium/uranium fuel cycle with recycling/fusion	++43,200,000	++90,200,000	Limitless	Limitless Rounded

*NEA Report 20.2 (Reference only – Added Recycle Technology (++).

**Rounded to 10,000 years

The 25,000 and limitless years are my projections which are very conservative.

CAVEAT

The 10,000 year projection is based on a 2%/yr growth. If the world accepts our plan, the growth in nuclear will be much more than 2%. Thats why the new energy sustainability council needs to be on top of this; and develop a graph charting: Sustainability Years vs: world conversion to nuclear. And provide energy direction for the world to use.

Nuclear fuel cost per pound (/lb.) has fallen from $1.28/lb. to $0.30/lb. 1980s to today

Energy generated with nuclear in 2012: 0.7 billion megawatts 2.5% increase from 2011

Resource (1000, T) (nuclear)
  Discovered   12,270
  Secondary Sources   910

  Unconventional resources-   4,039,110

Conventional resources could last ~25,000 years with a usage rate of 2,000 t/year
Source: EIA

    The uranium supplier, Cameco, reported on its website today (8/16/14) that the market today for uranium is 170 mm lb., and will increase to 240 mm lb. this decade, with ninety-one new reactors worldwide.
    Japan has reported to the NRC that it will be restarting nineteen of its shuttered reactors after Fukushima.

13

How Do We Plan for a Sustainable and Cheap Energy Future?

"Sustainable" is the first important word in this chapter's title. The second is "cheap." There are all other sorts of ways to produce expensive energy. The best way to understand sustainable is to compare it to a subject we all understand—our everyday economics/finances. If we have an income of $2,000 per month, expenses of $4,000, and with savings of $50,000, we know that this situation is unsustainable for a long time. That's synonymous with where we are in our energy fuels. We are consuming a lot of our existing resources without using much of our sustainable (renewable) resources. So how do we develop a plan that puts us on a sustainable energy path?

It's similar to the example above—we can outline a plan to stabilize the finances above. This may include reducing expenses to $3,000 per month and get a second job making $1,500 per month and saving $500 per month to add to our $50,000 savings.

Energy is a little more complicated, but the world plan outlined herein will solve the problem. It will not be easy. There are many challenges. The major hurdles are education for the public and politicians, and truth in the media and speech. This book starts the education process. The second is to elect politicians that are willing to study the subject thoroughly and act on recommendations of the "experts" in the field. (The second challenge also means that our politicians *not* politicize the matter. Just like the debt and unemployment, our children/grandchildren and future generations are depending on us. Time is of the essence before we run out of oil and other FF, along with destroying the atmosphere that all living species of fauna and flora needed to sustain life.

So therefore, with the answers to the basic questions of cost (Chapter 11) and resource supplies (Chapters 9 and 12), we have our answers.

Some of these may be covered in other chapters, so bear with me. I'm in a hurry to get our book published.
1. Electricity generation cost: Nuclear is the cheapest among renewables. The sun and wind are expensive, and their harvestable supplies are limited.
2. Resources - supplies: Nuclear has supplies for 25,000 years compared to oil at 30 years and coal/gas at 200 years.

The obvious answer to our energy needs is nuclear power (Chapter 12).

Speaking of sustainability, if we have evolved enough to have a sustainability council for wild seafood (e.g, the Marine Stewardship Council, a standards and regulations group), why is it that private industries haven't done the same for energy?

We know the answer—FF. industries don't want a collective regulating their activities or worst yet, sponsoring an energy resource that is cheaper than FF. The Marine Council doesn't have this problem. Wild seafood are owned by the public. This has led us to our final conclusion; we recommend the governments of the world via the United Nations establish a subgroup labeled the World Energy Sustainability Council.

The United States should lead the world in this endeavor for the following reasons:

1. We are the biggest economy in the world.
2. We consume the most oil and FF resources per capita.
3. Most important, we have the most advanced democracy in the world. Maybe this will show the world that "democracy can deliver on energy as well."
4. We have the most advanced reactor technology. This will be squandered if future administrations don't fund nuclear research projects.
5. Others will follow our lead, especially after we demonstrate to the world our willingness to change the course of history with our energy plan.
6. President Obama will be unemployed in 2016.

Our collective has the potential to unite the world, bringing together Eastern and Western countries to not only save our planet but also persist with cheap power (2C) for many generations to come. We will have plenty of volunteers (better than not having a job).

Is there any other issue that has a chance to unite the world? You know the answer.

14

Nuclear Power Around the World

Reported by the Nuclear Energy Institute (NEI) and the DOE Energy Information Administration (EIA), other news sources and websites

According to the International Atomic Energy Agency's database, nuclear power provided 12.3% of the world's electricity production in 2011. Thirteen countries rely on nuclear energy to supply at least 25% of their total power needs

As of 2013, thirty-one countries operate 434 reactors, with 71 plants mostly under construction in fourteen countries. The United States has five of these. China has 30, Russia and India have 11 and seven, respectively. Each plant will generate about 1,000 MWe. This shows how far behind the United States is relative to these Eastern countries, especially China. This is a rueful state for the United States to be in, as the inventor of nuclear power and the best technology for reactors in the world (Westinghouse).

These 435 power reactors have a capacity of 370 G MWe, about 13% of world energy capacity. There are 71 reactors under construction or almost so, mostly in the Asian region, as indicated below (revised):

China - 34
USA - 3
Russia - 13 including one in Ukraine
India - 7
Iran - 1

Cameco, America's largest supplier of uranium, reported in its website that 92 new reactors will be added within the decade. Read Chapter 14.7 about China's global nuclear plans and how it has already built 26 of these NPP around the world, utilizing Westinghouse and Russian technology, and Russia, with the largest reactors at 2,000 MWe capacity, is expanding India's 2,000 MWe at Kudankulam and building 10 new NPP over the next 10 years.

14.1

Great Britain

Britain is expanding its interest in nuclear power, with the building of the Hinkley Point B plant in southwest England. The plant will supply power for over five million homes and supply 7% of Britain's energy needs at a cost of $22 billion. This fits into Britain's plan to build 12 new plants by 2030. The French will build the plant, and China will be the majority owner. This expands China's interest by having access to France's technology. China has plans to build 10 new nuclear plants in the next 10 years. China already exceeds the United States in nuclear-generating capacity at 1,145 gigawatts compared to 1,000 gigawatts.

In China. this represents only 1% of country's generating capacity compared to 10% in the United States. (NYT, 10/2/2014) Britain has 15 existing nuclear plants that are operated by the French. From this, we can see that the United States is letting other countries beat us to the business. This is due to our policies on nuclear energy. There's no other reason! Except our focus on shutting down the government! Read-on, it gets worst!

14.2

Germany

Merkel declared that Germany will close all nuclear plants in the next few years. The chancellor made that decision after Fukushima (Japan) and Germany has spent billions to install renewable energy plants. Since then, the price of energy has skyrocketed, causing Germany to not only subsidize the investments but also to now subsidize energy prices for residents and companies. If Germany stays on this path, we forecast it will lose its competitiveness (9/1/13).

And just this month (10/13), Obama proposed a plan for the United States to spend another $8 billion to build renewable energy power replacements. NPP can be built without this much of a subsidy, and the power is cheaper.

A Genergicbilanzen, an association that tracks energy projects in Germany reported that this country has converted 23% of its electricity needs from nuclear to renewables at a cost of over $100 billion, and will spend up to $550 billion to complete its plan to reach 80% conversion to renewables. This exposes what can happen if the public guides their politicians based on EV fears and lies. Germany is the most productive and competitive country in Europe. Once Germany completes the conversion, its products (cars, etc.) will no longer be competitive.

One cannot compete with 10C electricity when others like the United States, Russia, China, and France are producing products using 2C electricity. Fear has so distorted Germany's thinking that although it has spent $100 billion on the NPP conversion, this country continues to raze existing villages so that mining companies can keep on extracting the coal under villages. Atterwasch is such a village (home to 25,000 people). And this is happening because Germany needs the coal until the conversion to renewables is complete. All because of the country's decision to shut down its NPP.

We are thankful the Obama administration is not as fearful as Merkel's administration. But be aware, it's the public's attitude on energy that drives these decisions in both countries. We have had a presidential candidate in the past who would have shuttered all U.S. NPP. So it's not a stretch to say that the United States could make the same mistake.

And it's not just Germany that is wobbling on strict sanctions on Russia regarding Ukraine/Crimea. Big Oil in the United States is applying pressure on Obama: because companies, like Exxon and Mobil, are heavily in Russian oil. These oil companies, like LUKoil, Rosneft, and Gazprom are owned by friends of Putin. Russia and the United States have a lot in common when it comes to who has power within their respective governments. Russia is a monarchy with oil controlling the president. The United States is an oligarchy with oil money controlling the government. Why else would taxpayers be funding the food-to-fuel programs?
Germany, WSJ, 8/27/14

Now the WSJ has caught up with what we have been predicting since 2011 (since Fukushima). The important numbers are:

1.0 Electricity prices paid by industry (not residential)

E per megawatt hr.

Nation	2007	2013	% Increase
Germany	88	125	+70%
China	85	70	-20%
France	53	65	+5%
USA	50	50	0%

E1 = $1.32

The WSJ got its information from IHS.

2.0 Germany's GDP decreased 0.6% last quarter.

3.0 It's a concern that Germany (and Europe) will lose its competitiveness going forward (our prediction in 2011).

4.0 BASF, the huge plastics company headquartered in Germany, uses more electricity than all of Denmark and employs more than 50,000 in Germany. BASF will not be investing more in Germany and may move some facilities elsewhere. Where will elsewhere be? Maybe in China, with its lower electricity price and its electricity prices decreasing by 20% over these five years.

5.0 In comparison, Germany's future electricity price is expected to go even higher as the country implements its renewable electricity program through 2025.

Note: We stumbled on this information looking for a WSJ report on the latest leaked IPCC reported in the NYT. I will continue reading these newspapers for more information.

But I'm not holding my breath until it comes. And now, the NYT, 10/27/14, reported that executives of BASF (Badische Anilin & Sodafabrik), a chemical giant with 33,000 employees in Germany (headquarters), are looking outside its base to expand primarily because of energy policies and cost.

See, just as we predicted, policies led by fear and a lack of knowledge can cause great harm to a country's future. Other countries around the world will capitalize on Germany's misguided direction. That's why, we are so concerned about democracies that have leaders who openly espouse their opinion that experts and science should be avoided. Without this work, we would never have stumbled on why the integration of a country's economy, energy, environment, and experts matters.

14.3

France

France *generates* 75% of its energy from nuclear and exports electricity worth three billion euro per year. France can do this because of its lower-cost nuclear electricity. And presently, France produces 17% of its electricity from recycled nuclear fuel.

France generates 542 billion kW with 58 nuclear reactors. In 2011, France made a decision to continue supporting nuclear plants. In 1974, France started its nuclear program primary because the country has little FF. France uses Westinghouse technology.

France also has some renewables: 28 gwa, 6.6 gwa, and 2.2 gwa from hydro, wind, and solar, respectively.

France is in this strong, powerful position because its citizens support nuclear power. 67% of the population believe climate protection is the best reason to go nuclear.

Early in 2008, France established the top level Nuclear Power Council (NPC). The head of the Atomic Energy Commission (AEC) is the secretary-general of defense and the chief of staff for NPC. So we can see that France aligns its policies on military, energy dependence, and economic development with the energy and environmental needs of the country. If France can achieve all this with very little FF, can you imagine what the world can achieve by going with our plan?

The *Pandora* show in Chapter 23 claims France should be the model for the world. We agree. France is the opposite of Germany. Even though France is much smaller than Germany, let's see how this saga plays out.

The latest news regarding the GE and Alstom joint venture was that it's blessed by France. The agreement gives the French government leverage to ensure employment in France and capitalize on the synergies in the energy sector regarding: renewable power plants and the electrical grid. Neither the NYT nor the WSJ mentioned nuclear. But you and I already know that France uses GE–Westinghouse nuclear technologies in its NPP. And we already know France is involved with China to help the British build their nuclear plants.

Talk about a collective. Ayn, goodbye to your philosophy against collectives for the poor; even capitalist use collectives to power their needs. It takes a big collective government or companies to build big projects of the world. And the United States better straighten out its problems. These other countries via their collective governments will be the leaders going into the next 50 centuries.

With separate readings, we can see that France's "socialist" government (a bad word in the United States) is making sure no French jobs will go overseas. This joint venture involves big capitalist companies like Mitsubishi and Siemens. Maybe France is ahead of the United States in more areas then nuclear and may even has a better constitution regarding allowing oligarchs to have such power.

We have not arrived at this port by being an ideologue. We arrived here by stumbling through life and experiencing its greatness. I'm no longer envious of elite educations; maybe just MIT engineering (and now economics).

We continue on with the U.S. "capitalism" permitting "individuals" like Mitt Romney to sell complete companies, including equipment technologies and jobs, to the Chinese, all to enhance his oligarch status. Enough already.

Another take: Look at all the animosity I have for protesters of the protestors during the Vietnam War, especially those that used their political position to easily get a draft deferment four times. My draft number was 72.

14.4

Argentina and Brazil

Argentina and Brazil have been cooperating on nuclear technologies and recently signed an agreement to build 30 reactors using recycled nuclear waste as its fuel.

The joint company that manages this program is EBEN. Currently, nuclear electricity generated in Brazil and Argentina are 3% and 7%, respectively. (12/10/13)

The 2030 energy plan for Brazil is to build four new plants with 1,000 mw each. Brazil presently gets 91% of its electricity from hydropower. By 2050, Brazil projects 60,000 mw nuclear energy with one new plant per year; 10 in 10 years.

Brazil will have to change its agreement with the United States before it can operate NPP that recycles waste. Waste recycling is not allowed in the United States. Read on and observe the dates.

New - Breaking News (NYT, 7/13/14)

Argentina and Russia signed contracts today. Russia will build the new nuclear reactors and provide nuclear technology for Argentina's new nuclear complex (reported above).

In addition, Russia will sign new nuclear agreements with Brazil shortly to expand Brazil's nuclear program. Russia will also be meeting with Venezuela, Bolivia, Uruguay, and Cuba this week. All of this in the face of the U.S. nuclear agreements that are already in place.

These dissatisfactions with the U.S. contracts facilitated these countries going with Russian technology (recycling and waste handling), and some funding (?). Again, the first country or region to get this right will dominate the world. Also, Russia has eliminated its EV opposition.

And we thought Crimea and Ukraine was our biggest worry with Russia (?).

14.5

Russia

Total electricity 25,000 MWe
18% nuclear
Plans to increase NPP from 31 to 60
A 100% increase as fast as possible.

By way of the World Nuclear Associations, Russia is participating in nuclear fusion reactors/technology. The country is allocating ~$50 billion dollars to nuclear development by 2015 (2 years). It will be increasing sales of Russian reactors and technology overseas. Argentina is an example.

I just looked today, 12/10/13, 12:00 p.m. Philippine time. Now we have Russia and China to complete with. Let's not have a self-inflicted tragedy by not funding our nuclear program. Our engineers and scientists can work with the world so that no one country controls the world's power. We sure are glad the United States had forward-looking politicians during the 1970s, 1980s, and 1990s. Without funding, the Soviet Union could beat us with this aspect of the Cold War. Also without funding, our greatness can be overcome. So there, Norquest and Ayn.

"Moore's Law"

Gordon Moore of Intel developed the theorem that the processing of microchips capacity will double every two years.

In Thomas Friedman's opinion article in the NYT on 3/26/14, he wrote about Putin's actions regarding Crimea and Ukraine and how Putin is challenging three powerful forces: mother nature, human nature, and Moore's law

First human nature: The Ukraine crisis started because the Ukrainian people wanted to leave the Russian monarchy and saddle up to the European wagon, and Putin is betting on FF at a time when the International Energy Agency has declared that two-thirds of FF will have to remain underground if the world is to limit climate change to a warming of only 2C. Friedman goes on to speculate that Moore's law will permit solar energy to displace FF energy before Putin's strategy can take hold. Friedman goes on to support America and Europe's push for more renewable energy.

In our opinion, Friedman doesn't understand (no knowledge) that Putin is already far ahead of the United States and Europe (except France). With Putin's push for cheap nuclear power, Russia has NPP already under construction in Ukraine, plus Russia has a plan to replace FF with nuclear. So if Friedman's speculation that Moore's Law applies to solar energy (and without any evidence that microchips have any relationship to solar technology), we think he has Moore's law confused with Murphy's law. In fact, Moore's Law is more applicable to nuclear's development.

Solar and wind are limited because of the limited areas suitable for harvesting of these fuels. In fact, we would go on to say that even if we harvested all the suitable wind and solar fuels in the world, this energy would not meet the energy needs of the world in 2100. Just do the math! And Friedman is so fearful of nuclear and EV that he doesn't even mention nuclear energy in his opinion. Now you tell me, what would cause this omission? Especially since Russia already has future plans to build a nuclear power program quickly.

Further in Ukraine, Westinghouse is extending its contract with Ukraine energy operator Energoatom to supply nuclear fuel for the latter's reactors. The deal is worth over $100 billion for five years. This will bolster Ukraine's commitment regarding long-term cooperation with the West. President Obama met with Ukraine Prime Minister Arseniy Yatsenyuk before meeting with Westinghouse managers. This example just shows how valuable it is for the United States to have the best nuclear technology in the world and to update our outdated treaties and contracts regarding nuclear fuel uses, like the treaties with South Korea, Argentina, and Brazil on the recycling of nuclear waste. An issue like this could prevent Western companies from getting contracts like the ones with Ukraine, Argentina, and others. Again, nuclear power for the world has so many advantages.

WSJ. 12/22/14

Russia will be expanding India's biggest NPP at Kudankulam with a 2,000 MWe reactor and will be supplying 10 new NPP over the next decade.

And now (4/24/15), the NYT reported that the United States is selling one-third of its uranium supplies to Russia. Again, Moniz's incompetence shows through. If the United States is truly pronuclear, this would not have happened. Moniz's thinking probably goes something like this "We have enough uranium to meet U.S. needs, so why not do the capitalistic thing and sell off these strategic resources." And Hillary Clinton believed her energy secretary. The money flow to Bill didn't hurt either. This reinforces our belief that the Clinton's are closet conservatives. The other Western countries, including Canada, that went along with the deal show us how the Eastern countries that believe in experts and govern from the top down make better decisions than Western countries that don't govern for the long term.

The deal puts Putin closer to his goal of controlling the world's uranium supply chain. Russia now controls 20% of our supplies, and Canada sold off some of its most productive mines in Kazakhstan.

This deal makes Rosatom the largest uranium producers in the world via its acquisition of Uranium One. Our book warns Western countries that "we better have a stake in the new technologies for atom energy or the technology will be controlled by Eastern countries." And now we find that the Eastern countries are preempting the West by controlling the uranium supply chain. And now I'm afraid again. So I will reemphasize our statement again: "We are still in a Cold War that we thought we won." Complacency and fuelish politicians can undo all the wars won by boots of soldiers on the ground.

And the information we are gathering gets even worst! The United States only produces 20% of our present uranium needs, so how will we build out our uranium plan without supplies? We will succumb to Russia's price no matter how high. Thanks, Hillary and Mr. Moniz.

14.6

United States

Energy Secretary Ernest Moniz is dragging his feet on nuclear power plant loans that were funded by Congress in 2005 with $17 billion approved at that time and were expected to expand to $50 billion in order to jumpstart a nuclear renaissance. So Moniz cannot blame anything except his leadership for not going forward. He is obviously guided by fear. He recently released $6.5 billion to Georgia Power for the first new nuclear power plant in three decades. This plant will be located near Augusta, Georgia and will be called the Vogtle nuclear power plant. It is close to the Savannah River research projects in Georgia.

In another demonstration plant for a nuclear enrichment project in Piketon, Ohio, in partnership with Constellation Energy, the company that owns Baltimore Gas and Electric, again Mr. Moniz is dragging his feet. And Mr. Moniz has killed many of the other nuclear research projects. It just goes to show the damage caused when driven by fear or *ignis fatuus*. Again, EV have been demonstrating against all of these new nuclear projects. If Moniz is afraid of these EVs, he should resign and go into hiding, and wait until Nader is president.

How could one man do so much destructive damage, to the world, in such a short time? Much shorter than Obama's time?

Other aspects of the U.S. position on nuclear power are spread throughout the book (not much though).

In the United States, the TVA is finally finishing the first nuclear power plant of the twenty-first century: the Watts Bar 2 Nuclear Power Plant. Commercial operations are expected to start next year (2015). This plant will displace some of the coal generation capacity at TVA. The government has spent only $4 billion to build and modernize the plant for efficiency and safety. This plant has provided good jobs for 3,000 workers since the early 1970s. The plant took such a long time because of stops and starts by politicians.

For example, President Carter appointed an antinuke political operative, David Freeman, to head the TVA in 1977. He called the executives of the agency "nuke-aholics," and he later wrote a book about his experiences at the agency. His book supports some of the EA's propaganda. These stops and starts caused the project to take more time to build than the building of the Panama Canal or the Great Pyramids of Cheops. A treasure this great is worth the wait. It's just too bad the world (with plenty of workers) can't move faster toward cheap energy and a clean environment. The workers and engineers on this project are ready to complete other plants like this one. The timeline for a project of this size should be more like five years. Being a professional engineer (PE) and project manager, just hire me to work with the Shaw Group or Bechtel. There are many PEs that can manage these projects. Another piece of information: This new plant is designed for a seismic 8 earthquake (upgraded after Fukushima). The evolution just keeps going.

The United States doesn't have a long-term plan for nuclear power because it doesn't have a long-term plan for energy.

14.7

China*

China currently has 20 reactors operating and has 34 mostly ready for construction. In order to provide 58 GWe by 2020 and 150 by 2030, China plans to "go global." It plans to sell to any country that wants the technology and equipment that China has available, (e.g., a combination of the technology the country has obtained from Russia and Western nations, plus China's own nuclear advancements).

China has Western technology that it continues to upgrade, including waste recycling, thorium, fast reactors, etc. This is the advantage China has—being able to select the best of both worlds.

China's plan is to install all nuclear for future growth and then to replace some FF plants with nuclear. China has some renewable power plants producing 400 GWe of clean energy, mostly from hydropower; wind provides 80 GWe and 3 GWe come from solar. China will add between five to 10 new nuclear reactors during its five-year planning cycle. And China will spend $120 billion by 2020 with $75 billion per year each year after that. China has plans for 10 GWe solar in the next few years.

*World Nuclear Association

And now we read about China offering up a referendum to make the selection of Hong Kong's chief executive *more democratic* by reducing procedural barriers, earlier, put in place by Beijing. Compare this to the voter depression tactics going on in the United States today.

China is acting on its long-term plan to "go global" with nuclear. Of the 71 new NPP being constructed globally, China is constructing 26. Westinghouse, a U.S. company is supplying China with its most advanced technology—the AP1000 reactor, with components from Curtiss-Wright Corp., another U.S. supplier. China is quickly building-out its capability by squeezing out other international suppliers, via technology, resources, and cheaper labor.

An example of this is the parts supplied by SPX Corp of Charlotte, North Carolina, for the AP1000 reactor that will be supplied and built by the Chinese state machinery company, the State Nuclear Power Technology Corp., a company under the control of China's central government. Westinghouse and other suppliers have signed technology transfer agreements with China, providing it with the technology for control of the total business. After all, once China has the nuclear business, why not build out all of the nuclear business. Western companies would do the same thing, except they cannot compete with a big government, Plus China uses the best experts and technology the world has developed. Now I'm scared! (WSJ 12/16/14)

A risk path going forward is a better choice than a safe path going backward.

China WSJ 5-16-15 (Update)

China will start eight new reactors this year, and it plans to expand nuclear power from 21 gigawatts to 58 gigawatts by 2020.

14.8

India*

Short term:

India expects to have 14,600 MWe nuclear power or 25% of requirements by 2050.

India is outside the Nuclear Proliferation Treaty. India has no indigenous uranium. Therefore, India is researching thorium for its nuclear needs. India's vision is to become a world leader in nuclear technology, especially for fast reactors and thorium fuel cycle. India consumed 774 billion kWh in 2011, doubling since 1990. Currently, India's nuclear power is only 3%. So for India to reach 25% of its energy needs with nuclear by 2050, it will have to build some 15 large nuclear plants every 10 years. India's electricity needs are expected to grow 6.8% every year.

Currently, about 33% of India's population is not connected to any electrical grid.

India will need $1,600 billion to support these plans through 2035.

*World Nuclear Association.

India is driven because it has very little FF supplies. India knows that uranium will cost less than 0.1% of what FF will cost in the future. India has no choice because of economics. India knows that its aggressive nuclear program will lead it out of poverty and will save the environment, if the world follows the India's plan.

India has this aggressive nuclear program even though it is not included in the nuclear nonproliferation treaty (NPT) wherein five countries are included. India was not included because of its nuclear weapons program in 1970. These five countries were awarded the status of nuclear weapons states under the NPT.

India has been granted cooperation agreements with the NPT. It's agreements like these that have allowed and continues to allow countries and companies to provide equipment and fuel for India's nuclear needs. Russia has offered a 30% discount for its ($2 billion) reactor. The Russian government is behind this effort. Westinghouse didn't get this business because the U.S. government is not supporting nuclear power worldwide.

India is just one example of what Russia has been doing with nuclear technologies and businesses. And we thought the Cold War was over. As of today (4/1/14), Russia is building NPP in the Ukraine and Crimea. Does nuclear power factor in to Russia's intrusion into these territories? What does it matter if they are owned by a monarch like Russia or an oligarch like the United States? It doesn't, as long as they bring with them nuclear support. At present, it looks like Russia is winning this part of the Cold War, and the United States doesn't even know we are still in the Cold War. The American public thinks the Cold War is over, and the United States and its allies won.

Long term:
WSJ, 12/22/14

India has contracted with Russia to expand its Kudankulam NPP with 2,000 MWe and to build 10 NPP within the next decade. This is just the start for India's very aggressive nuclear program. How will India meet its carbon commitments? You and I know the answer!

14.9

South Korea

Being a democratic country and an ally of the United States, South Korea has nuclear power technology, and the Korea Electric Power Corp. (KEPCO) is trying to build more nuclear power plants so that South Korea can export electricity mostly to North Korea and others. However, KEPCO is currently fighting landowners who don't want power lines passing over their burial grounds. This is probably a guise for opposition to nuclear power by EVs. Either way, it shows the basic fight between ancient and modern times. We realize this should be an easy decision. What could be better than selling inexpensive electricity to a country that desperately needs electricity. Maybe, if South Korea could deliver inexpensive electricity to North Korea, the latter would slow down its drive for nuclear technology. What are the negatives? None, except an ancient culture that hasn't been in a position to educate the public (knowledge again).

According to the World Nuclear Association, Korea generates 30% of its electricity from nuclear power with 23 reactors generating 20.7 GWe and is planning to expand to 40 reactors for 56% electricity needs. In addition, South Korea is exporting nuclear power technology to other places in the world. This country recently won a $20 billion contract to build a plant in the United Arab Emirates.

Korea imports 97% of FF needs for a bill over $170 billion; nuclear is its choice for the future. Sounds familiar. Korea is trying to get additional contracts with Turkey, Jordan, Romania, as well as East Asia. This brings up a point. U.S. government restrictions on companies providing technology and plants to foreign countries, should be renewed and relaxed immediately. U.S. companies cannot be competitive with these restrictions. The NPP will be built no matter what! These might as well be built by U.S. companies and maybe with U.S. government assistance rather than its imposed restrictions. Another reason to fund the Export-Import Bank, republicans.

14.10

Japan

Japan has decided to shut down all nuclear power plants by 2020. One of its first projects is to build a wind farm off the coast of Fukushima, 12 miles from the closed nuclear plants. The goal is to build 140 wind turbines by 2020. This farm is projected to generate 1 gigawatt of electricity, equivalent to the power generated from one of the closed NPP. This is an ideal location since the transmission lines are already available.

The Japanese government will allocate $226 million (tax dollars) for three turbines. The total project will cost over $10 billion. A new nuclear plant would cost maybe $5–8 billion.

According to a spokesman for Shimizu, one of the consortium of companies involved with the project, this is Japan's only hope without NPP. A couple of positives are the following:

1. 100% Japanese employment
2. The money will go back into the Japanese economy.

We pray the United States never finds itself in a bind like this. We better elect smarter politicians because many of the one's we have don't even believe in science, much less what scientists and engineers can achieve.

Japan imports 84% of its power fuel. With Japan shutting down many nuclear plants after Fukushima, Japan's cost for FF replacement will be 3.6 trillion yen in 2014. Let's hope the country can survive this hit to an already weak economy. After the Fukushima wind turbines come on line in 2016, the electricity prices will be even higher, in the meantime pumping a zillion tons of anticlimate emissions into the atmosphere.

Prior to Fukushima, Japan's plan was to increase nuclear power from 30% (50 GWe.) to 40% by 2017, with a goal to increase energy self-sufficiency to 70% by 2030, all by replacing FF plants with NPP. This was expected to reduce CO_2 emissions by 70%. *Source:* Japan Atomic Energy Agency (JAEA)

Japan has a spent fuel recycling program, including enrichment and reprocessing of spent fuels for recycling. Again, this is against the law in the United States and one of the complaints Korea has with its Western agreements.

And today (1/15/14), Japan announced it will be starting up some of the closed nuclear plants. Sooner or later, the truth (lower energy prices) will rise to the surface. We hope to be alive when this happens.

Of the 48 nuclear plants that were idled after Fukushima, one of the Ikata nuclear reactors will be restarted this summer. None of the 48 reactors were damaged as a result of the magnitude 7.5 earthquake and the worst tsunami in Japanese history it triggered. The reactors were shut down because of fear; fear can be a variable that can destroy the best work or investments for any situation.

Prime Minister Shinzo Abe has the backing of the Liberal Democratic Party and their close ties with the nuclear industry. That's how Abe was able to restart some of the reactors. Abe's success so far has not been easy. Japan has the same problem with misinformation from EV and Big Oil that the United States has. Yeats had it right: "The best lack conviction, while the worst are full of passionate intensity." So, all the world has to do is to bring out the compassion and passion that is latent in the best of us.

In Japan, Abe has this backing because the country is falling even further behind due to the expensive FF. And once the wind projects come online, the public will bear the burden of very expensive wind projects. I thought the Japanese were smarter. The fear factor again.

Japan Update (NYT, 2/12/14)

Japan continues to keep most of its NPP online. The recent election of Tokyo's governor shows how close Japan is coming to making a full decision to shut down all of its NPP. Although two antinuclear EA were defeated, they only lost by about 200,000 votes. Abe was emboldened with the win and promised to restart some of the idled NPP. Abe faces an unwilling public without a national consensus. With the polls showing an ambivalence on a scenario without the NPP, Japan will have to purchase FF costing $360 billion over the next 10 years.

While the business community vehemently supports nuclear energy, they understand the need to have cheap energy in order to continue Abe's renewed economic uplift. Without the NPP, the business community knows Japan cannot compete globally with countries like United States, China, France, and others. Japan is having the same problem as the United States, the public hears and believes the lies/fear voiced by the EVs. And our governments are afraid to take a bold stand. Japan, like the United States, is weakened because the public is not behind nuclear. Abe may be ahead of Obama with his latest statements.

(3/16/14)

Japan announced it will be starting up some of the closed nuclear plants.

Latest update, 8/15/14

Cameco, a supplier of uranium, reported that Japan has submitted applications to the NRC to start up 19 reactors. Look, already Abe is influencing the public against the EVs. Let's hope he can continue, and restart all 48 reactors.

14.11

Iran*

Iran has one commercial nuclear reactor, the Bushehr Power Plant. Russia has a tentative agreement to take uranium stockpiled in Iran, and convert it into specialized fuel rods. The agreement hinges on Iran reaching agreements with the West. Russia has helped the West reach these agreements. And in fact, an Iran–Russia agreement fits in with the need to get some of the 28,000 pounds of uranium out of the country.

Source: International Atomic Energy Agency (IAEA)

The reactor is a heavy-water type with plutonium in it spends waste. Both Russia and the West have the same objective: to cut off every pathway to keep Iran from developing a nuclear weapon. Iran doesn't trust Russia any more than it trusts the United States and the West. The negotiation partners with Iran include the United States, Britain, France, Russia, and China. With an agreement to send the uranium to Russia, this opens up the negotiations so that Iran can keep its 19,000 centrifuges operational.

*World Nuclear Association

The one plant in Iran produces 240 billion kWh, with a goal of 23,000 MWe so that oil and gas can be freed up for sale. Iran has agreements with Siemens KWU (German) and Framatome (Frencia) for these new plants.

These plants will use Russian technology (primarily), with German technology for upgrades to the reactors. The plant in the Bushehr province started in 1975. The plant is 24% German, 40% Russian, and the balance Iranian. The plant will be operated with a 50–50 Russian joint venture and with a WER-1000 (MWe reactor). Two more plants are planned for the site. These plants will be designed for a seismic 8 rating. Iran is planning for six more NPP in 15 years and two desalination plants using the 2C electricity. Iran is working with the IAEA for an LWR (light water reactor) with 360 MWe capacity (Westinghouse technology). One Bushehr reactor will free up $2 billion per year of oil and gas and with no anticlimate emissions.

With Iran's nuclear plans for the future, they are way ahead of many other countries. Now (WSJ, 11/12/14), Iran is reported acting on the plans to carry out its nuclear program with new Russian agreements (in addition to fuel processing agreements). This calls for the nuclear company Rosaton to build two more reactor units at the country's Russian-built Bushehr plant and four additional reactor units at another site to be named later. Iran claims these agreements expand "our cooperation for decades to come." These new nuclear plants will be built and operated under safeguards mandated by the International Atomic Energy Agency. The fuel will be produced in Russia. The spent fuel containing "plutonium" will be sent back to Russia for reprocessing to recover the radioactive materials. These materials will be "recycled." Russia built the first reactors at Bushehr in 2011 and turned the plant over to Iran in 2013.

Iran already exports about 10% of its electricity, with plans to export more to Middle Eastern Countries that don't have a 2C program.

14.12

Hungary

Hungary has a vibrant nuclear program with 46% of its electricity generated via nuclear and with plans to expand nuclear to 60% within the next decade. Hungary has agreements with the West, but it also has agreements with Russia to provide fuel for its Russian-built reactors and to send some of its spent waste back to Russia for recycling (once again, the recycling issue with Western agreements).

Weaknesses: Hungary

Hungary is an example of weaknesses with Western democratic capitalism. Mr. Orban, the prime minister, led Hungary away from communism with the collapse of the Soviet Union 25 years ago. And now Hungary is a member of NATO and the European Union.

Mr. Orban wants Hungary and other Eastern European countries such as Romania, Bulgaria, and others to reevaluate Western ways and to consider going the way of an authoritarian liberal democracy like China, Turkey, and Singapore.

Mr. Orban cited the Western governments' failure to anticipate and deal with the economic collapse of 2008 and the ensuing recession. Mr. Orban went on to say, "Western democracies will probably be incapable of maintaining their global competiveness, without owning up to their problems and the willingness to change themselves significantly."

Mr. Orban expressed his concern that although democracy has brought freedoms, it has also brought loss of jobs, inequality, and a feeling of insecurity.

The world is a big place, and much work is taking place to make human life as comfortable as possible. Maybe it's time for the world to organize itself around a huge collective, with maybe Obama as our leader! (Once he reads this book and changes his mind on the nuclear issue.) After all, he is guided by his energy secretary, Mr. Muniz, and maybe by a little fear. With this knowledge, I'm sure Michelle will change his mind.

14.13

Africa

Installed generation capacity in South Africa is 54.7 GWe, 5% nuclear. Recently, it signed contracts with Russia to build eight WER Russian reactors which will supply 9,600 megawatts. This capacity will be used to replace coal generation and to supply future electrical needs. The financing of $27 billion is in the final stages with Rosatom, the nuclear power company in Russia. Rosatom has contracts for 29 NPP with China, India, Turkey, Vietnam, Finland, and Hungary.

China is lining up contracts for $93 billion to build the first six Russian reactors by 2030. The United States was involved via a consortium with Westinghouse and the Shaw Group. They had to back out because the U.S. Congress cut funding for smaller NPP via German technology. China is proceeding via Tsinghua University and two Chinese companies both run by the government—China Nuclear Power Corp. and Nuclear Power China General. Both are funded 95% by the government. The Westinghouse consortium has been working on smaller NPP of 100 MWe to 200 MWe.

These smaller NPP are important to all of Africa because its countries can't afford the larger NPP investments. Mr. Gates via TerraPower is also involved with this research and prototype development. The German technology for these small NPP modules are built with twin reactors and a single turbine that could be built offshore and transported to the site.

Now we realize the nonproliferation of nuclear materials is an issue that affects these decisions, but if this causes a slowdown for cheap power for the world, these considerations need to be rethought. Another example of the weaknesses currently in Western democracies. Hungary is the other example.

If I had enough time and funding, I'm sure there are many more. Other than the bigger countries, Africa and Hungary are the only countries I researched, and we discovered problems with each of these semi-Western democracies. Argentina and Brazil also have problems with the nuclear agreements imposed by the Western democracies. Iran doesn't have these problems yet!

Because of the remoteness of areas in Africa, most Africans do not have access to grid electricity. This limits Africa's ability to expand their GNP.

Economist (6-6-15)

Many Africans have access to electricity using FF Generators. These generators cost 50 cents/Kwh and sometimes up to $10.00/Kwh to operate.

Other than nuclear, Africa has many rivers, such as the Congo River System, that has the capacity to generate 40,000 Mw, twenty times the capacity of Hoover Dam. As reported in 11.8, dams with capacities larger than 4.0 Mw can generate 2C electricity (the only fuel source that can do this other than nuclear).

See how much hydropower China has built with their river capacities (14.3). The World Bank is studying how to finance these projects and their environmental impact. Even with Africa's electricity situation, they have made commitments to eliminate deforestation by 2030. Let's all hope Africa can go forward with either nuclear or hydropower to build an electrical grid system so that their public has access to 2C power.

Note: HP will flood thousands of acres, NPP will take-up less than 20 acres. Let's all pray Africans can make a good decision going forward.

14.14

Turkey

Electricity production in Turkey is 240 billion kWh, with FF imports of $60 billion in 2012. The country has tried to start up several nuclear plants since 1992, but it had to stop because of the lack of government guarantees. In 2008, Turkey signed agreements with IAEA. Turkey's cooperation agreements were signed with the United States and China in 2008. Then in April 2009, agreements were signed with Rosatom (Russia), with the first reactor to start up in 2016 and plans to build eight NPP by 2026 Russia will be the primary owner by supplying $20 billion for the $22 billion project. Russia will be the operator and the fuel supplier. Long-term plans are to generate 100 GWe by 2030. Spent fuel will be repatriated (recycled) to Russia. This is in line with the normal practice for Russian-built plants in non-nuclear-weapon states.

15.0

Closed and Open Nuclear Power Systems

The closed nuclear power system refers to the recycling and reuse of nuclear spent waste products. This system has the advantages of:
1. Reducing the waste materials
2. Recovering the economic value of the spend uranium in the waste

The United States and Western partners (except France) do not recycle because of security reasons. Eastern countries such as Russia and China do some recycling. With the technology today, the nuclear spent waste could be reduced by 20%. Research is continuing to develop closed systems so that the spent waste can be further reduced and the economic value recovered. The research is being carried out by the U.S. government, international governments and private companies, like TerraPower and Westinghouse. TerraPower is developing the technology for small NPP to be used mostly by remote, poorer areas. Laws need to be changed so that these areas can then have access to 2C power.

Congress and the president are behind the eight ball when it comes to these out-of-date laws and theirs effects on our Western allies regarding our treaties and agreements with them. The United States is losing a lot of business because of these laws and fear.

This nuclear research is referred to as the fourth generation nuclear reactor. The nuclear industry (Western) supports the open system. Eastern countries support the closed system. See Argentina, Hungary, and Germany (Chapter 14) for the effect Western laws are having on all Western businesses, both private and governmental. China and Russia are taking advantage of our government's fuelishness and out-of-date laws. China and Russia know it can move faster than the West because:

1. The country doesn't have to educate the public.
2. And this is a big one—They believe and will execute the science of the experts.

16.0

The Tail-end of Nuclear Power: Nuclear Spent Fuel

There is no research here yet except the importance for the 250,000-year supply situation regarding recycling and recovery of the spent fuel value. Eastern countries support research to further extend the science. TerraPower and Westinghouse are two companies in the United States that are developing spent-fuel recycling technology. Our government should help. Since Obama's administration is wasting so much money, anyway, we might as well waste some on waste projects.

Added at last minute, 11/26/13 2:00 p.m., Philippine Time

On the last comment, I know later I will need to get into the Yucca Mountain project. That place is so toxic I may have to wear a lead suit to get in. At least, I will have the suit for later. I should not be making fun of such a serious issue.

Yucca Mountain has been financed by taxpayers for 30 years, and in 2010, it was shut down by Energy Secretary Moniz. The Government Accountability Office issued a report saying the shutdown was political and would set back safe waste storage for decades. The Obama administration provided no scientific reasons or any assessment of the risks associated with the hasty closure. And now, nuclear operators are suing the Obama administration and are losing every suit. This has cost taxpayers billions of dollars and delayed the clean-up of FF emissions for decades. Thank you, EVs, Muniz, and the Congress. The election is coming up, let's clean-house (including the Senate!).

Also, once the recycling technology is completely finished, all the waste currently stored can be recycled. Yucca may not be needed.

Speaking of waste, what about the problems we have with the spent waste from coal fired power plants? Just this week, Governor McCrory of North Carolina used his office to negotiate a deal with

Duke Energy. Duke will not have to clean up the recent huge spill, some 73 Olympic-size pools of lagoon waste that spilled into the Pan River in North Carolina. This was after the huge-spill in Tennessee. The governor negotiated the deal because of the governor's previous position as manager and owner of Duke Energy and Duke's sponsoring of McCrory for governor.

There are thousands of lagoons like these in the United States, and some 30 are in North Carolina where Duke is located. This spill in North Carolina has contaminated thousands of acres and tens of thousands of people along the rivers effected by the Pan River spill. In order to clean up the whole mess, the federal government with taxpayer funding will have to step in. Who else can do this once Duke goes bankrupt? *Another bail-out!*

17

Updated Global Energy Outlook, 2014

Energy Information Administration (EIA)

This forecast takes into account the recent gas and shale oil discoveries and the recent technology to extract using fracturing. World energy consumption will grow by 56% between 2011 and 2040 from 524 quadrillion Btu in 2010 to 820 quadrillion in 2040. With this energy increase, carbon emissions will increase from 31 billion metric tons to 45 metric tons per year by 2040. This projection is supported by Shell and Exxon.

The updated conclusion regarding the recent supplies of gas and shale oil is:

"This *gas and oil* production growth is important, providing some breathing room. Nevertheless, the projections by pundits and some government agencies that these technologies can provide endless growth and lead to a new era of energy independence are entirely unwarranted based on the fundamentals. At the end of the day, fossil fuels are finite, etc." These newer supplies do not change the conclusions *we* reached based on our forecast in 2012. Even a very optimistic forecast will only increase oil supplies for another 10 years. So what, we run out in 40 years instead of 30 years! Our goal is 25,000 years, short term and 250,000 years, long term.

Another point for the United States is this: "Even with these production increases, we are still importing 40% of our oil needs."

18

Smaller Power Plants

Smaller plants are plants with less than 1,000 MWe.

Large plants - 1,000 MWe
Small plants - 25 to 200 MWe
Mini - to 25 MWe
Micro - to 200 homes

The world energy organizations, like IAEA, WNA, and the USA-DOE, are researching and engineering these plants at a very rapid pace. So even if I could reach conclusions/recommendations today. Tomorrow the results would be invalid.

Plant Size		Plant Cost (in $)	
Large	-	5B	20B
Medium	-	1B	5B
Small	-	100M	1B
Mini	-	1.0M	100M

In the United States, some of the larger companies working with the government are:

1. Westinghouse
2. McDermott
3. Bechtel and other large engineering/construction companies, like the Shaw Group
4. Babcock Wilcox
5. General Electric
6. TerraPower
7. General Dynamics

Table 18-1
Power Cost (Smaller Power Plants)

Fuel	Generations Cost cents/kWh
Nuclear	2.0*
Wind	No technologies available (to match 2.0)
Coal	" " " "
Gas	" " " "
Oil	" " " "
Solar	" " " "

*Technology, 100% complete; engineering and prototype development, 20% to 61% complete. (Once complete, the cost will be between 2 and 10.) All can be completed by 2025, with adequate government funding. Or maybe the Hungary Project with Russia funding will lead the way for smaller NPP.

Note: The United States, via Westinghouse and the Shaw Group, had to drop out of these projects after the U.S. Congress cut the funding. The United States had already invested many billions of dollars before Congress stopped the funding. Thank you, Tea Party.

Nuclear is the only fuel that can generate 2c power with no climate emissions, except hydropower in rivers that are large enough to generate at least the size of Hoover Dam.

Just another dream! And to think that we are already 50% of the way there. Being a scientist/engineer and project manager, I can assure you this is no dream.

So let's wake up the world to this reality. (Some are almost awake.)

Sorry, I can't spend any more time on this aspect of our research. After all, I'm just a retired, disabled engineer living out of my van and using Starbucks as my office.

Engineers and scientists in big companies, with their government's backing, *can* accomplish these goals within the next couple of decades. We need two things: public support, and compassionate and passionate leadership. Obama has both of these qualities; he just needs to evolve more. He has already evolved on other issues. We just need to wake him up.

The objective of all of this work on smaller plants is to bring 2C nuclear power to areas that can't afford the investment for large NPP.

Today, the target is to complete the work by 2025. Let's not let the EV and FF industries stop us from going forward.

Motto

A risk path going forward is a better choice than a safe-path going backward.

So now we have a motto and label for the nuclear renaissance.
The label is appropriate for energy and the environment: 2C.

Don't be confused by the smaller power plant label. Micropower plants that can service up to 200 residences might be quite viable in isolated communities, like Alaska, remote islands, and the African bush. They can serve as sun and wind replacements for oil generators with oil cost at $100 per mm Btu. The cost per kWh, in comparison, is about $1.00 per kWh for smaller power plants. The investment to install these systems can be as little as several hundred thousand dollars. Sun and wind energy is certainly better than burning FF. The same can be said for individual homes and water heating systems. Again, this is better than FF-fired generators.

But don't be beguiled into thinking these systems are cost-effective compared to large or smaller power plants supplying electricity to the grid. The economics of these microsystems have already been proven for other utilities. If you ever lived in a house that has a well or a cistern, you would know that "when the grid is available," one almost always chooses to connect to the grid. The reason is that the maintenance and fuel cost are in the range of $1.00 kWh. The worst case price for larger and smaller power systems is less than $0.1/kWh compared to the $1.00//kWh.

Many places in the world cannot afford the several thousand dollars for power that cost $1.00/kWh. This is the reason two billion people do not have access to electricity and why Bill Gates works on smaller nuclear plants that will greatly benefit the world. He will go down in history as the greatest philanthropist ever, leaving a much-better legacy than the greatest computer person ever.

19.

Nuclear Power - Environmental Issues

As stated elsewhere and as is known generally by the public, and in case there are still some that don't know, nuclear power plants, just like wind and solar, do not give off any, that's 0 *zero* harmful climate emissions. When one passes a power plant, the plume one sees is 100% water vapor. Another interesting point is that FF plants waste about 33% of the energy given off when the FF is burned. Furthermore, all this heat is taken up by the cooling systems. And in some FF plants, the heat in the water creates problems with fish being sucked up with the cooling water, and the water discharge temperature can disrupt the ecosystem. NPP use cooling water to control the systems.

Otherwise, the only environmental issue is the waste (e.g., spent nuclear materials) as addressed in Chapter 16.

This book doesn't go into detail regarding the damaging climate change caused by CO_2 and other anticlimate emissions released from burning FF either for power or transportation. It's interesting to note that atmospheric CO_2 levels track the burning of FF. Prior to the Industrial Revolution, CO_2 levels were consistent at about 250 ppm. After 150 years of burning 50% of FF resources, the CO_2 level has almost doubled to 400 ppm, and we are on course to reach 500 ppm shortly. And we predict the level to hit 800 ppm after burning all FF resources at the rate we are burning today. As we discover more FF resources, the CO_2 level will be even higher.

By developing a plan to extend FF for 25,000 years with this burn rate, the environmental CO_2 level will stabilize at current levels. This essentially means that the environmental problems we are experiencing today will not get worse, except in the ocean levels that are not in equilibrium today. As we (the world) get smarter managing world energy supplies, we could start to reduce CO_2 levels, and consequently, the climate will get better. It is scary to even imagine the worst-case scenario. Our world plan accomplishes the twofold 2C goals covering the environment and energy.

I started this book on energy and power, thinking, I didn't need to write much about the environmental issue, since so much information is available. However, I continued reading in many of our premier newspapers that some of our politicians don't believe climate change is a *net* negative for the world.

Evidence: Just today (11/5/13) in the WSJ, Bret Stephens tried to make the claim that climate change is a positive (net) for the world because in some areas in the North, climate change is a benefit. Mr. Stephens goes on to quote from *Notes on the State of Virginia* written by Thomas Jefferson and published in 1785.

This is how nonscientist can greatly mislead the public by taking the positive from the oeuvre provided by a scientist.

In his article "Does Environmentalism Cause Amnesia?" Mr. Stephens tried to make a case that adverse climate change as seen by the UN's IPCC report is similar to the neo-Malthusian forecast of the 1960s and 1970s when journalists and other fiction writers were forecasting the population growth would outstrip the world's capability to feed the population. So this comparison, plus the *Notes on the State of Virginia* by Jefferson, are supposed to debunk the IPCC report. How ridiculous! He went on to point out the positive effects explained in the report, trying to explain that the scientists should have used these causes to explain that climate change is good for the world! His following assertions show fuelishness at its best:

1. A warming climate seems to correlate positively with greater food production.
2. The positive effects that CO_2 has on photosynthesis.

The fiction writer went on to make the case that growth would outstrip food production. This gave us such titles as *Famine 1975!*, a 1967 bestseller by the brothers William and Paul Paddock, along with Paul Ehrlich's vastly influential *The Population Bomb*, a book that began with the words, "The battle to feed all of humanity is over. In the 1970s and 1980s hundreds of millions of people will starve to death in spite of any crash programs embarked upon now."

In case you're wondering what happened with that battle to feed humanity, the UN's Food and Agriculture Organization has some useful figures on its website. In 1968, the year Mr. Ehrlich's book first appeared, Asia produced 46,321,114 tons of maize and 439,579,934 tons of cereals. By 2011, the respective figures had risen to 270,316,205 tons, up 484%, and 1,289,633,254 tons, up 193%.

It's the same story nearly everywhere one looks. In Africa, maize production was up 247% between 1968 and 2011, while production of so-called primary vegetables has risen 319%; in South America, it's 308% and 199%. Meanwhile, the world's population rose to just under seven billion from about 3.7 billion, an increase of about 90%. The population is predicted to rise by another 33% by 2050.

What about the supposedly warming climate? According to the EPA, "average temperatures have risen more quickly since the late 1970s," with the contiguous 48 states warming "faster than the global rate." Yet, U.S. food production over the same time has also risen by robust percentages even as the number of acres under cultivation has been steadily falling for decades.

In other words, even if you believe the temperature records, a warming climate seems to correlate positively with greater food production. This has mainly to do with better farming practices and the widespread introduction of genetically modified (GMO) crops, and perhaps also the stimulating effects that carbon dioxide has on photosynthesis (though this is debated). Warming also could mean that northern latitudes, now not suited for farming, might become so in the future.

The world isn't likely to be getting any hungrier. Quite the opposite: Purely natural (as opposed to man-made) famines are becoming unknown. As the Irish economist Cormac Ó Gráda noted in a 2010 paper: "In global terms, the margin over subsistence is now much wider than it was a generation ago." This also holds for former famine zones such as India and Bangladesh, whereas China, once the "land of famine," nowadays faces a growing problem of childhood obesity." Only in Africa is food scarcity still an issue. But even there recent food crises in Malawi and Niger did not result in major loss of life.

What *does* hurt people is bad public policy. One is the U.S. food-to-fuel mandate—justified in part as a response to global warming—which diverts the corn crop to fuel production and sent global food prices soaring in 2008. Another is the cult of organic farming and knee-jerk opposition to GMOs (genetically modified crops), which risk depriving farmers in poor countries of high-yield, nutrient-rich crops. A third is the effort to ban DDT without adequate substitutes to stop the spread of malaria. Malaria kills nearly 900,000 people, mostly children in sub-Saharan Africa, alone, with each passing year. The list goes on and on. With these issues, we agree 100%. But these bad public policies are disconnected from the warming issue.

Environmentalists tend to have conveniently short memories, especially when it comes to their own mistakes. They would do better to learn from history. Just take the quote about the warming climate with which this column began. It's from *Notes on the State of Virginia* by Thomas Jefferson, published in 1785.

Being a scientist/engineer, I don't understand how the writings and thinking of Thomas Jefferson in 1785 could be used as a justification for not believing the expert scientists of today.

The only possible reason is that Mr. Stephens needed to support his ideology. The UN's IPCC report clearly concluded that the negative 2% on food supplies is after including the positive benefits Mr. Stephens pointed out in his article. And the WSJ never reported directly on the IPCC report.

The U.S. newspaper reported today, 11/4/13, that food production will start falling in 2020. (Information reported by the EIA)

Mr. Yelgin, vice chairman of IHS and author of *The Quest: Energy Security and the Remaking of the Modern World*, focused on the U.S. efforts to become independent of imported oil. He concluded that although the United States has had this goal for a long time. We are still importing 35% of consumption. The same as was the case 40 years ago. Mr. Yelgin also concluded that if the world continues on this path, FF will be depleted in the relatively near future. Our prediction is 30 years for oil.

Herein, we have stated that the world has depleted about 50% of the known FF supplies since the beginning of the Industrial Revolution. What's more alarming is that 50% of this consumption has occurred in the last 20 years.

Mr. Jonathan Lynn, a spokesman for IPCC, did suggest that it's not too late to have a positive impact, and that many countries are adapting to climate change with most of the funding for projects to take place in the next decade, with results expected in the late twenty-first century. Obama signed an executive order on 11/1/13 to step up efforts. Another report will be released in Berlin in April (2014). This report will analyze potential methods to limit the rise of greenhouse gases. If it comes out before I finish my research on this book I will include it herein. Guess what? It's covered.

"There's a lot of scientific and engineering work that the world is funding on this crisis. Let's take advantage of it and act. We are leaders of the world." (Mr. Bloomberg the ex-mayor on New York)

As explained in the NYT on 7/15/13:

Mayor Bloomberg brought managerial acumen to New York that can't be matched by others. Bloomberg is an engineer at heart. He studied engineering and brought his set of skills to city hall. "It is an approach to the world, a strategic sensibility different from a politician's. Engineering favors innovative solutions over incremental fixes, 'calculation over consensus.'" I could not have stated it better myself.

If I can sit in Magic Johnson's Starbucks (south central Los Angeles, California) with my laptop and can function as a retired, disabled engineer, then there are companies and big governments that can change the world. But the public must support and push them. I love my children, grandchildren, and future generations, and want the very best for them. Let's leave this world in better shape than when we entered it.

The NYT just this morning (Saturday 11/2/13) reported on a draft report from the IPCC, that was leaked early. It's not final, but it follows up on the oeuvre already available on this subject. This report is more inauspicious than the report issued in 2007

With quantified results regarding *food* supplies decreasing by 2% already, with animals and other species trying to survive the *new* climate. So what's going to happen if CO_2 level are allowed to increase from 500 ppm to 800 ppm. Yes, "allowed." We can change now. Again, that's why today, I have made a decision to hurry up and finish this book, and at the same time, I will release $10,000 of my retirement money to secure an advertising spot with the NYT and *Readers Digest*, my publisher's suggestion. I'm planning on living to be 120, so I will need a return on these investments. So if you are reading this book, thanks so much. Now back to the topic.

Food *demand* is projected to grow by 14% during this decade. The IPCC scientists won the 2007 Nobel Peace Prize for their work. And just like this book, we scientists and engineers may not have the exact numbers on the projections correct, but just like other big subjects, we might be a little off, or we could be a lot more reasonable than what's actually going to happen.

Scientists have reported that some areas will benefit from the climate change. This is what scientist do. We analyze the positives along with the negatives and report on these sequiturs but never the apogee or nadir—just a balance. Leave it to the politicians and fiction writers to pick a side and choose either, especially when it comes to discrediting scientists and engineers. It is the scientist's job to interpret the data; it's not for the politicians, rich aristocrats or fiction political writers. And we should believe the experts. The media reports too many of the latter, confusing the public with "what the truth is."

The CO_2 levels will stabilize! The problem is what will this stabilization level be? (See Chapter 11) The quicker we start our "world plan" the lower the stabilization level will be.

The latest news we have from NYT, 12/1/14, on the Peru IPCC Summit this week is that an agreement has been reached on the temperature increase target of 2C. Even realizing the 2C target is too low, the scientists must compromise so that a final agreement will be signed at the 2015 conference in Paris. The scientists issued their standard plea: The world is going to hell fast, and our efforts to limit the rise to 2C cannot be realized because the world is not reacting fast enough. UN negotiators can't give up. Any agreement will limit the ultimate damage and prevent the worst-case scenario (800 ppm CO_2). No matter the amount, all delays are costly.

The problem the IPPC is having is this:

"The oceans are not in equilibrium with the 2C climate today, and there's not enough big data to scientifically postulate the final outcome. This might lead to beguiling cataclysmic results even if 2C is achieved." And the scientists don't want to have a target that ends up with this result. They would much rather have a lower 2C target. Ice sheets in Greenland and Antarctica will continue melting for many years with the damage already done. Not only will the melting sheets cause the oceans to rise higher than predicted, but more and more CO_2 will be released to the atmosphere because of the trapped CO_2 in the ice. This dilemma can be solved with the power of 2C electricity; start replacing FF and build new power plants with nuclear and some renewables. Target temperatures will not be needed.

Richard B. Alley, a climate scientist at Pennsylvania State University, predicted that with the 2C target, oceans will rise 23 feet, supported by calculations of climate scientist at the University of Florida, where Andrea Dutton, a research chemist studies sea levels around the world. NYT, 12/1/14

Lima, Peru, United Nations summit to convene this week with thousands of diplomats, to forge an agreement on anticlimate change emissions so that the agreement can be signed in 2015. Scientists issued their standard plea: The world is going to hell fast, and our efforts to limit the temperature rise to 3.6°F, even with the agreement, cannot be obtained because the world is not reacting fast enough. UN negotiators should not give up because any agreement will reduce the ultimate damage coming and already here, and will now just try to prevent the worst-case scenario: an uninhabitable world. The National Oceanic and Atmospheric Administration will be reporting 2014 will be warmest year on record. Let's see if the WSJ reports on this gathering and science reports issued. So far none.

Russian oligarch and billionaire G. Timchenko, is lobbying the U.S. Senate with $280,000+, plus $320,000+ by others (all friends of Putin) to minimize sanctions over Ukraine and Crimea. (WSJ, 12/1/14) What about the Peru Summit? Okay, an opinion article appeared the next day. However, it only mentioned the summit a couple of times, then used the introduction to explain how Obama is using climate change to ultimately take over the electricity industry, just like what he did with health care, explaining how this will be a winning issue for Republicans because of:

1. Fewer jobs
2. Expensive energy
3. Indefinite commitment to pay billions for climate aid

And we say "bring it on."

Our research on false and misleading reporting on energy is only one example of what's occurring in the WSJ vs. NYT scenario. Reporting

from the media's CIA coverage is another example, far worse than our detailed energy study. Here's just a quick look at the differences:
NYT, 12/10/14
1. Quotes from the CIA report on the front page of the NYT
2. Cites how Dick Cheney authorized the CIA to withhold information
3. How CIA leaked classified information (a crime) to Mr. Kessler, and he wrote a book *The CIA at War* (based on the leaked material) without the information that was withheld by Cheney and the CIA.

WSJ, 12/10/14
1. Report written by Democrats without Republicans on the panel
2. The paper picks and chooses some positives from the report that proves torture works, contrary to the sequitur of the CIA report.
3. The CIA fights back and "still claims that torture saved lives."

To believe this assessment of the report, one would have to believe U.S. democracy is corrupt to the core. Is this another example of the anti-expert culture some media's espouses? After all, five years and $40 million develop a certain level of expertise on the subject, and the issue can be debated by claiming the "science" is not settled. We don't know. Enough, our expertise is energy!

One more thought to contemplate. Which media is more influenced by oligarchs than the other? And which media is "always" claiming they are the fair-and-balanced folks?
NYT 12-16-14

This ocean rise of 23 feet may happen soon, or it may take 1,000 years, nobody knows. So what does this mean for us? The world is mobilizing around the 2C target, realizing that this is not low enough to prevent the worst case of 23 feet ocean rise, and the storms and droughts worse than what we are seeing today.

Our conclusion is that the IPCC cannot handle the problem without the help from other energy groups around the world. This, plus our work on energy sustainability and cost, has led us to one of our recommendations: form a new world group to not only address the energy environmental issue but also to address energy sustainability.

This new group should be within the United Nations and called the *United Nations Energy Sustainability Council*. This concept is covered more thoroughly elsewhere in this book; that's how we came up with the bifurcated 2C label for energy and the environment. The combination of these two UN groups will be powerful enough to move the world toward a sustainable and emission free, 2C environment.

If we let the CO_2 levels increase to 800 ppm, climate scientists tell us that human suffering will be great, with the world average temperature average of 10 °F more than it is today. Can we survive these levels? No one knows. Thanks to Ford, I have a floating shelter! And I will need it in Watts, California. Many poor people live in these lower levels.

Let's proceed with evaluating the reporting on the IPCC reports and what the media has presented to the public. The IPCC reported on September 27, 2013, from Stockholm, the sixth report since 1990. Read on, now we have the seventh report, each with greater certainty that humans are the major cause, and that if we want to keep the temperature rise below 2C, countries will have to *stop* burning fossil fuels now.

This book proves that the only way to do this and keep energy prices low is to build new nuclear power plants for future growth and then start replacing FF plants as soon as possible.

Exxon Mobil commented on the IPCC report "because of shareholder pressure," claiming the value of its reserves will not be affected because:
1. The world will need 35% more energy in 2040 than today, in line with our previous projections herein.
2. To meet the goal of 80% reduction in anticlimate emissions will cost the average U.S. household an additional $2,350 a year in energy expenses, and governments don't have the will power to impose these limits.
3. Renewables cannot make up the energy needs, and therefore, FF will be needed at an ever increasing price.

Exxon Mobil acknowledges to its shareholders that climate change is occurring and the world needs to reduce anticlimate gas emissions. Its solutions are the following:
1. Conservation

2. Allow domestic companies to export their supplies of national gas to the world because national gas reduces anticlimate gases compared to coal or oil. Plus, the company can get higher prices overseas.

One can read into this report that Exxon Mobil is primarily interested in keeping the price of oil and gas at the highest price possible. Exxon never discussed the use of nuclear energy to make up the energy needs. One certainly can't blame the company for its position. It's the company's job to satisfy its shareholders.

My comment: "Exxon Mobil knows it can hire an army of EVs to keep nuclear at bay."

The Exxon Mobil's comments on the IPCC report can be found in Exxon's website. The WSJ has not covered the climate report other than the short section reported above. The WSJ knows it made a mistake with its last opinion on the previous climate reports. And it must realize it is on the wrong side with this matter.

NYT, 1/15/14

IPCC recommends a tax on carbon emissions. "Some Republicans question the underlying science," and with lawmakers tied to the FF industry, and with the rise of the tea party in recent years, a tax increase is unlikely." This highlights the problem the Republicans have with science and the problem Democrats have with the science of nuclear power.

The NYT article goes on and on about the evidence of climate change problems, plus the likely policy changes that need to happen before the existing laws, plus a carbon tax, can have any chance for emissions to be reduced. On and on, but the NYT doesn't say anything about "nuclear being the logical way to reduce carbon emissions." But at least it is reporting the news on the matter.

Where is the WSJ on this matter? Just plain fear is the only logical answer for its behavior. And our sequitur highlights the importance of living a fearless life.

The WSJ finally on 4/10/14 had an article on the IPCC report, but only reporting on the recommendation to reduce greenhouse gases to keep the temperature from rising above 2C. There was no mention of all the other effects of climate warming, only citing critics' responses. The WSJ had a perfect opportunity to push for an aggressive nuclear power program to stop all anticlimate gases. On the same day, the WSJ had an article about California's push for greater renewable energy use and how the EAs are causing the renewable energy to be more expensive. Regarding environmental group lawsuits, the WSJ has a point here.

So finally in finishing this research, the WSJ published an article 4/27/14 on the IPCC report. Don't get excited yet; the article was in a weekend edition, written by a British zoologist and philosopher turned sci-fi writer, Mr. Matt Ridley, educated at Oxford. He is a conservative member of the House of Commons in Britain. He has written many conservative articles. He wrote a weekly column for the WSJ called "Mind and Matter." Not our matter.

Mr. Ridley has published many fiction books. Mr. Ridley's philosophy "is that the human race can innovate its way out of problems without any planning." So as a writer, he essentially writes fiction around his ideology regarding anticlimate gases. He believes humans caused most of the emissions. At least, he knows that much. He disagrees with the projections of doom. He believes humans can innovate and utilize the emissions for useful purposes. He believes "the emissions so far have produced more good than harm," a belief that is contrary to the climate panel report. He goes on and on with his fiction, no science or facts, just good fiction from a fiction political writer.

The WSJ article is titled "A World Running on Full." He uses the same type of fictional writing to create an ideological story around how to innovate our way out. In some ways, we can agree. But scientist and engineers never bet on new unproven technology, only after the technology is proven and workable in at least a prototype environment. And it takes planning to lead us to the solution. Mr. Ridley's job for a number of years was the manager of the Blagdon Hall Estate after his father died. When asked what the most dangerous matter he fears, he answered, "The government is the problem rather than the solution." That's what all these oligarchs across the world believe because governments are the only entity big enough to prevent the oligarchs from enslaving the world.

What kind of fiction do you think Ridley will weave to show that the iron law of economics produces more good than bad results. I'm sure he can, and it will be a top seller with the WSJ sponsoring it.

Since I don't know anything about the House of Lords in Great Britain, I did a little research and discovered that members are supported by taxpayers. Also, according to the NYT 8-23-15 and to Mr. Kirsty Blackman, a member of the Scottish National Party and recently elected to the House of Commons: "The House is bloated, outdated, undemocratic and a 'horrendous waste of money.'" Others go on to say members should be "better vetted."

And we say major newspapers should "better vet" their opinion writers. On the other hand, the WSJ knew about Mr. Riley's major conflict of interest but hired him to write his opinion pieces anyway.

Does this lead us to "a need for the government to better regulate our unfettered free speech?" You know the answer. All of this leads us to the reason why so many people are climate deniers in the West. It's not the fuelishness of individual writers, but rather our system of governance. The Evolution Revolution will correct this problem, along with the supreme court and congress problems.

Mr. Riley is a House of Lords member (inherited).

No wonder George Washington fought so hard to defeat the British kings and oligarchs. And we expect better news reporting than to use a philosophical fiction writers to analyze and draw conclusions based on fiction. This is especially true when the work of the expert scientists is before us. In many ways, using a fiction writer to report on this matter is publishing "a fiction story around an agenda." How can we expect anything other than a collection of misinformation for the public to decipher? The educational level needed to decipher and encode this matter gets to be higher and higher. Before a democratic system can work, we must step up to the plate and make sure all children and adults have access to this higher level of education. The one big advantage autocrats have over free people is that: "The public doesn't have to agree with expert science. The public needs to be better educated to analyze and decipher the truth from misinformation and storytelling."

So now we have to wait to see what fictional story Ridley will weave regarding Piketty's economic science that proves the iron law of the free market system, that is, the rich get richer, and his warning that big data suggest that Western countries, especially the United States, is "drifting toward oligarch." Now really, it's nice to confirm the law, but middle-class Americans see this as nothing more than the elites proving what's already "common sense" or learning by wandering around. This iron law says that big governments are the only way to get some of this inequality to trickle down and to do this with progressive taxation policies.

One more Bret article in the WSJ.
WSJ, 1/13/15, "The Scandal of Free Speech"

Bret equates QUIP (Queers United in Power) to the Al Qaeda group that killed Charlie. The main point he wants to make is that QUIP wants to use its collective strength to terrorize the world. How ridiculous. And I'm not queer, just a lover of queers. On the same day the WSJ had this article, "Obama's Dead-End Community College Plan." Obviously, the article puts down Obama's plan. I'm a graduate of Pensacola Junior College before transferring to Auburn. Prior to PJC, Auburn would not accept me.

I'm sure Bret, Matt and Rubert are nice people, they just lack the knowledge acquired via being a peripatetic graduate,—without an agenda, just the knowledge of the truth, as determined by experts in the field.

And now after our criticism of M. Riley's climate change denier claims, we discovered why Mr. Riley supports the deniers. In his editorial/opinion article in the WSJ dated August 14, 2015, it was disclosed that Mr. Riley is a land baron in England, inherited from his father. He gets his riches from the mining of coal on a land he and his family own. Here's the issue: Why would a worldwide media outlet like the WSJ hire a writer who has such conflict of interest to express the views of the paper? Shouldn't this be against the canons of good journalism? In today's article he reported that a Mr. Kevin Dayaratan of the Heritage Foundation, analyzed EIA data and concluded that Obama's plan to reduce carbon emissions by 32% will cost the economy one trillion dollars. The point here, other than the conflict of interest in reporting, is how can Mr. Dayaratan possibly calculate the effect on GNP without knowing how Obama's plan will be implemented. For example, if carbon emission reductions come from building new nuclear plants or other renewables, the method has a great impact on GNP. And what about the downstream impact of converting the cost of electricity from fuel that costs $ 3.0/mm BTU to a cost of 2 cents? The major disruption in GNP is the reduction of monies going to King Coal and the payments for carbon emissions. And Riley reports on this without any knowledge regarding energy/fuels. Another article by Riley shows up later (WSJ 9/5/14).

There's a new book on the subject, *935 Lies,* by Mr. Lewis. He wrote about "the decline of America's moral integrity." Mr. Lewis is well qualified to author this book being the founder of the Center for Public Integrity in the 1980s. He launched institutional nonprofit journalism in America. He wrote about the 935 lies that for-profit journalism has fostered and supported via the money interest of for-profit media. These include "well-connected power people" (oligarchs) "and companies with questionable policies and practices," and the journalists don't have any incentive to investigate before reporting. And many of these lies were used to support our involvement in the Iraq war. He went on to write that "serious journalism" should be undertaken with "great caution," especially when going for a mass audience that is not in the interest of owners. He asserted that this is coupled with the illusion that investigative reporting is more profitable than "the real thing." He blamed the decline mostly on "short-sighted greed and increasing corporatization."

Amazing! The Freedom House book brings the matter together (i.e., the same sequiturs herein related to the nuclear matter and the same sequiturs related to economic matters discovered by Piketty et al.) all because experts tilted away until the observation could be converted into theory. And then on to predictions and actions. All in accordance with Newton's mechanics and Galileo's work with apples.

I'm trying to complete this book, but the saga just keeps going forward. Every day new information supports our findings and world plan. I have read the NYT, WSJ, and LAT every day for five years, in order to have the latest information available for you. I will continue for a little while longer.

In addition to Ridley's climate change fictional writing, he also disagreed with all the scientists of the world, even Big Oil scientists, on the issue "that the world is rapidly using up the FF and the world shouldn't spend money for future generations" (even as a sign in San Francisco reads, "What has future generations done for us recently?").

Being an engineer and large-project manager, I have firsthand experience that a philosophy like Mr. Ridley's would lead to a total disaster when applied to big projects or programs. Innovations are great, but any engineer (project manager) that relies on new technologies without any proven prototypes would put his reputation, job, and company at great risk. If people don't plan for the worst-case scenario, they can be left with their pants down to their knees.

Again, it's Murphy's law. Just look at what happened with ObamaCare. These project and program examples can be applied to anticlimate gas emissions. No scientist of any standing would ever go forward without a plan to solve the problem. It's as simple as "The first step in solving a problem is to recognize that the problem exists." Sci-fi writing has no place in scientific communities. And after we discovered that Ridley has a major conflict of interest with his WSJ writings, he not only writes sci-fi; he has a financial interest in being a climate denier.
5/7/14
Both the NYT and WSJ had articles regarding the IPCC Report.

WSJ - Three columns on the fourth page, emphasizing "this may lead to taxes on businesses." Then an editorial blaming higher food prices on the FED, p. 19.

NYT - Story on pages 1 and 2 then page 13, emphasizing the projections of a 10 °F increase in average temperatures and oceans rising up to six feet. The report moves the problem from the future to the present, with the public still skeptical. The 23-foot. ocean rise was predicted later, so go back and review the dates.

Neither paper said anything about nuclear power being the only way to save the planet and provide cheaper power.

Just one last addition, 5/14/14. Two independent science groups reported in the *American Journal of Science* and *Geophysical Research Letters* regarding how they came to the same anticlimate conclusions but with different means. Scientist Thomas P. Wagner who runs NASA's programs on polar ice research reported that the two West Antarctic glaciers have melted enough to set off an inherent instability in the sheet, causing the sheets to retreat to deeper sections of the ocean floors, causing faster melting. And it's already too late to keep this from happening. Also, further warming will destabilize the Antarctic and Greenland glaciers. When this event is complete, ocean levels will rise to as much as 10 feet, four feet more than what the first IPCC report predicted. And already, Florida is having problems with the Atlantic seeping into homes. And this is happening today with only a 10-inch ocean rise near Miami. I'm sure happy to be living in my amphibian Ford 150 Van.

And yesterday, Senator Rubio from Florida, a Republican candidate for president, came out agreeing with the Republican platform that the climate is always changing, and there's not enough evidence that humans are causing the problem. He must be reading only the WSJ.

Then there's the book by Erik Conway, *The Collapse of Western Civilization: A View From the Future*. This book inspired Naomi Oreskes to quit her work in oceanography. As an historian of science at Harvard, she researched 1,000 articles in peer-reviewed scientific literature on climate change. These articles appeared over the last 10 years. She then published her book, *Merchants of Doubt: How a Handful of Scientists Obscured the Truth on Issues from Tobacco Smoke to Global Warming*. She discovered like we have herewith: that the battle wasn't science but economics. And the deniers all believe that unfettered free markets can solve or cure all problems. So the science of issues, like climate change and tobacco smoke, must be fought on the science to prevent attacks on the foundations of our democracy. Because any "thinking person" knows that the only way to correct the "harm" caused by unfettered capitalism is by correcting "unfettered" with regulations. And one of the best ways to do this is with a campaign of obfuscation about science and scientists. These deniers learned this from the successful campaigns financed by Big Tobacco for 50 years. So now Ms. Oreskes is accused of trying to destroy capitalism, etc. Now we know why Stephens, et al want to discredit experts and big governments that finance projects for the public good.

When we think long enough, it all comes together. By the way, of the 1,000 articles reviewed by Ms. Oreskes. Her sequitur was that no author had any evidence refuting the findings of the IPCC. All of this confirms our findings on the weaknesses of our capitalistic democracy system. Before we can expect to be successful against the autocrats and oligarchs of the world, as President Orbán from Hungary stated, "Extreme changes in Western democracies need to be made" (see Hungary, Chapter 14).

Oligarchs

These oligarchs all operate the same way, whether its textbooks in Russia or bridges in the United States harming the citizenry to help wealthy friends. All humans are equal, but some are more equal than others, just like in the textbook *Animal Farm*. Both examples show how the oligarchs ignore the experts and campaign on the premise that "the science is not settled" and then get members of the oligarchs to write up sci-fi on the subject.

Once Einstein established the theory of relativity, $E = Mc^2$, and mainstream scientists confirmed the theory, those that had a minority theory $E=Mc$ had no position in science. The reason is because science is science.

Jenkins needs to get back to business reporting where minority opinions have a position; or maybe even turn to writing about economics and politics, but stay away from science. At least with Ridley, he writes good sci-fi.

And look at what's happened to radio stations after the oligarchs took over the biggest stations in the United States. Back in the good old days, stations like the big one in Cincinnati, Ohio, WLW, broadcast across half the country with 50,000 watts power. I would listen to country music like it was the only station to penetrate through the Appalachian Hills. Now WLW broadcasts the hate speech concerning the ad hominem/feminnum that still resides in people who can't get over the South losing to the North. Why would this last so long? Because these monologues have no balance to the hate speech. We see this in Mr. Roof's killing nine people at the AME church in Charleston SC. Let me give you an example of what they get away with. In 2008, when Hillary Clinton was running for president, WLW was broadcasting one of the hate monologues, and the subject was Ms. Clinton's support of gay rights. The monologuist went to say something what I thought could never be spoken on radio, even with the untrammeled right to free speech they all praise. Anyway, I heard this:

"Hillary Clinton has eaten more 'pussy' than Bill Clinton has." Some of the hate speech monologuists include Bill (Willie) Cunningham, Savage, Limbaugh, and Hannity.

And now back to Mr. Stephens, WSJ, 11/7/14, and his political views on the Ebola crisis. I reference this because it clearly shows why he writes the way he does. He doesn't have any use for scientists, engineers, or experts, including the groups (collectives) it takes to research, build, construct, and code from a scientific basis. No other basis works. Some of these groups, in addition to scientists and engineers are designers, secretaries, teachers, professors, doctors, and many, many more. How can sci-fi writers like him disparage all these groups? They can because they have media outlets like the WSJ and Fox that are funded by the oligarchs. And they're supposed to be journalists reporting on facts. Where do they think these facts come from? Any lay person walking the streets using his/her sense perception can report on facts perceived.

So Mr. Stephens, WSJ, 11/7/14, believes: "Government bureaucracy should be treated, at every level, as inherently and inescapably incompetent." And that "expert opinion should be viewed as mistaken until proven otherwise." Mr. Stephens uses this opinion to support his view that the Obama administration should ban travel to countries that have Ebola patients, including Liberia, Guinea, and Sierra Leone. He disagrees with the "experts" that believe banning travel creates more of a problem, etc. We certainly don't disagree that he has a right to his opinion, but our beef is just on his starting point—"experts are inescapably incompetent." He further cites a century worth of experience going back to the progressive administration of Woodrow Wilson. To our knowledge, the United States during this past century has been a democracy. Maybe he was thinking of China. No, that couldn't happen because China has risen to be the second most powerful nation in the world with its expertly guided government.

And now, 9/25/14, NYT: Apiaries in twenty-two states are experiencing CCD (colony collapse disorder) resulting in mass die-off of worker bees. Bees have three enemies: diseases, chemicals, and habitat interruptions. Today, the United States is losing 30% of its bee population each year. The author of this information is Noah Wilson-Rich in his book, *The Bee: A National History*. Mr. Rich and the NYT don't try to explain which of these enemies cause the most destruction of bees but only say "we need to start doing things differently." We conclude the one thing we can easily do different is to stop destroying the atmosphere with anticlimate emissions. Just like birds, bees, and other very vulnerable species they are the first to die off, when will our time come?

And the LAT today (9/23/14 reported on a research claiming Northwest warming and droughts are not caused by human activities but by natural shifts in ocean winds. The findings were published in the *Journal Proceedings of the National Academy of Sciences*. The research studied conditions from the 1900s to 2012. The authors claimed their model doesn't show CO_2 levels correlations to the warming conditions. The LAT hasn't reported much on climate change then published this research. Neither does it report that other scientists, such as Kerry A. Emanuel, a renowned climate scientist from MIT, stated these findings are not inconsistent with what experts (IPCC) know about climate change. "Common belief is that it's globally, consistent, it isn't." There are even a few places over a couple of hundred years that have actually cooled, such as the southeastern United States (NYT same date).

These analysis and reports show us all how difficult it is to get to the *truth* about a subject. This is especially true regarding climate change or nuclear power (e.g., one of the weaknesses of democracy regarding policies established by the voting public using bad information rather than policies by experts in the field). Again, this is just a problem, not a condition that cannot be overcome.

All of us in California hope the extreme drought and hot weather is not caused by human activities. If this is true, we have a chance to survive the water shortages and floods. Otherwise, the "as we know it" prediction will come true.

The current gross domestic product (GDP) of the United States is $17 trillion. If China grows at its current rate, China's GDP will increase by $51.0 trillion in 2033, far surpassing the United States. Maybe a government by experts has economic advantages. Democratic societies will need to address their weaknesses before China uses its progress to militarize the world. We are convinced, this is accomplishable by the West, but we must get on the risk path to 2C power quickly.

The WSJ opinion, 10/1/14, published the view of Mr. Jenkins Jr., titled "Google's Climate Name-Calling." This opinion demonstrates the shallowness of the media to write something that supports their agenda. When Mr. Schmidt, Google's chairman, answered a question regarding support for ALEX, he stated that they joined ALEX originally because it supports business-friendly policies. Google recently resigned from ALEX because of its support of "climate deniers." Mr. Jenkins went on to rant against Mr. Schmidt for not understanding ALEX's position on the issue, plus all the reasons why "deniers" are important to the debate. Jenkins went on and on about the importance of deniers, even going to the extent of comparing deniers with the "crucifixion of an apostate by dinner companions." He went on to claim that "institutions of science are so mired in advocacy on the topic of dangerous anthropogenic climate change that the checks and balances in science, particularly with regard to minority perspectives, are broken." Jenkins's material came from a climate researcher, Judith Curry at Georgia Tech. Mr. Jenkins must have searched far and wide to find such a statement in order to support the WSJ position, etc. Mr. Jenkins has no background in science, much less how science actually works. To believe that science has a minority position contrary to "the proven theorems of the science" just shows how little Mr. Jenkins knows about science.

These climate change deniers are not playing stupid because they are stupid, but rather because to acknowledge the truth means they will be acknowledging their responsibility and expense that comes with a society that hasn't been cleaning up its mess all along. Another reason is they find a media outlet that allows them to write opinions even if the outlet knows they have a financial interest in the outcome, not paying to clean up the mess, that is, not disclosing the conflict issue until later. Matt Ridley, WSJ 9/5/14
"Whatever happened to global warming?"

At least this article is better written than the previous one. He reported that "most science journalists are strongly biased in favor of reporting alarming predictions, rather than neutral facts." He goes on to support his opinions from other like-minded journalists who extract misleading data and write their own opinions or calculations to support their agenda. An example is the Statistical calculations by Ross McKitrick, a professor of economics at the University of Guelph in Canada. He tried to prove "the burst of warming that preceded the millennium lasted about 20 years and was preceded by 30 years of slight cooling after 1940, and concluded that "the world will probably be only 1 degree Celsius warmer in 2100 than today, rather than the IPCC latest report trying to limit the rise to 2C. Ridley continues to write his fiction based on spurious opinions that favor his agenda. He never once acknowledges the IPCC leaked report found in the NYT. This leaked report reinforces last year's IPCC report that we need to "act fast" to keep the temperature rise to less than 2C. The leaked report upgraded the ocean rise to 23 feet. And after which the largest world association of scientists agreed (reported here in above). This is another example of confusing opinion writers that try to support an agenda rather than write the truth, as reported by the experts in the field. It's just too easy to write fiction when supported by a parable media outlet. Is Rubert involved here?

Our prediction came true;

The WSJ on 9/5/14 had the fiction writer "Ridley" report his opinion as the opinion of the WSJ. One especially important statement he made was that "all the other science journalists are biased regarding 'natural facts.'" These natural facts he cited are nothing more than the "sense perception" that has made it difficult for science to be understood by the general public, and especially when the sci-fi writers conjure up stories that play up to the sense perception. The *flat world* is a good example, I hope other journalists pick up on this and run Matt out of the community of writers.

These articles today were published before the 8/14/15 article where it was disclosed in the WSJ that Matt has a financial interest in being a climate denier. His family gets riches from digging coal from their land in Northern Britain.

Evidence:

The EV are the villains, which is why we created a special name for the EA. They lie and are criminals (the EV). We don't want to tarnish the EA. We appreciate some of their work regarding

1. Endangered species
2. Protection of our waterways, etc.

But EA need to be critical of their actions when the overall harm is greater than the good. Recently EV, for the second time, on 8/15/14, attacked an energy distribution station in California. This didn't hurt the power company; this damage is paid for by the public. And any disruption in service hurts the public. The morality of this is on par with 9/11. A better choice would be to lay down in front of the big trucks and don't get up until the point is made too late. Isn't this what the terrorist did in 9/11? This would not cost the public any money, except maybe a funeral.

8/29/14 - Looking for information from WSJ on the recent leaked IPCC report. Nothing yet. But the paper did have an opinion article titled, "Threats Bigger than Climate Change," because Kerry said he would be a passionate advocate for CC." So the six threats Kerry should be focusing on are the following:

1. Enemies around the world are intent on harming the United States.
2. Iraq is a greater challenge.
3. The Afghanistan "the administration has lost the will to fight."
4. Russia lacks the resolve to confront.
5. Iran should stop negotiating indefinitely.
6. Syria needs more U.S. assistance.

We all agree these six threats are very important to the United States. No comments needed. The scientists/engineers have it right. With the bent the WSJ has toward the Republican agenda, I'm sure they would agree that several of these problems can be solved with three or four more wars. All of this supports our need to be critical of the bias that exists in our media, one of the three legs supporting our democracy.

One more: the greatest evidence so far regarding Australia's heat wave in 2013-2014.

Five research groups using different climate models and big data science came to the same conclusion—the extreme heat wave was caused by human activities such as anticlimate emissions (NYT, 9/30/14). The new data not only pointed to Australia but to also heat waves in Europe, China, Japan, Korea and California. Each conclusion was the same. Martin P. Hoerling, a climate scientist with the National Oceanic and Atmospheric Administration, in the past has been skeptical of the links with human activity. Now he goes on to say, "The evidences in the latest research reports is very strong." More research is continuing, with new reports in the next couple of months.

I've said this before, but I promise this will be my last comment on climate change. This comment is important because Obama is planning to become a leader of the world of nations, with a summit on 9/22/14 called by the United Nations in New York. The summit has a single goal: to build momentum for a deal on climate change by 2015.

The WSJ this week had an article on "deforestation," saying it doesn't harm climate change and maybe even helping to cool the environment. All the world's panel scientists must have missed this relationship. I didn't report this earlier because it doesn't fit with our conclusions!

Africa plans to announce at the summit that it promises to halve deforestation by 2020 and stop all deforestation by 2030. If Africa can make such a contribution, the least the world can do is to electrify all of Africa with nuclear power plants, or hydropower (Chapter 14).

John Podesta, a consultant for Mr. Obama, said Obama is taking the summit seriously and plans to lead these leaders of 160 countries with proposals for: (1) helping the most vulnerable and (2) a plan to culminate in Paris in 2015. He will be challenging others to step up to the plate, with additional funding for the Green Climate Fund, which has been in place for many years but only has $1billion dollars from Germany. So for Obama to be successful, he knows he needs the public to put pressure on Congress. I believe Obama can get the public behind him once he reads about our plan. All it takes is a fearless speech on the benefits of nuclear power around the world. The Chinese, Russians, and Indians already are planning on nuclear power to meet their commitments on reducing CO_2 emissions. The world leaders will be left with the problem of what to do with the mess we have already made. But stopping the new emissions now is the first thing to do.

Also in support of this summit, protests around the world will take place this week. This may get to be bigger and bigger, but we will stop here. Just a quick look at the controversy over the carbon rule. All of this doesn't *matter* to us because if our world plan were to be implemented, the carbon problem is solved. The Republicans, via Issa et al., are suing the EPA because they believe the Natural Resources Defense Council (NRCD) inappropriately influenced the EPA in writing the carbon tax legislation. The article is in the NYT, 10/11/14, p. A17. We don't much care about this issue except that it's wasting taxpayers' money. And the NRCD wants agreements that help the United States with government commitments. What a waste on both sides, Republicans and Democrats. This happens when a country is afraid to "rock the boat." And EA need to realize that it is acting without much knowledge (my comment only). For us, nothing else needs to be said.

The summit on 9/22/14 cited above occurred. Obama and China agreed on cutting back on carbon emissions.

N/Y/T/ 11/3/14:

The IPCC issued another warning regarding global warming. The agency is doing its best to get the message out to the world in the most severe language to date. As summarized in the NYT: "Without immediate action, flooding, food shortages, and mass extinctions are likely."

WSJ, no report on this today. I will continue looking and will report herein if something arrives.

Scientists from the national Audubon Society reported climate change is causing great disruptions in the habitat and death. All the way from the Baltimore orioles to the California raptors and to the brown pelican on the Gulf Coast, many bird species are endangered because of droughts and heat. The three-toed woodpecker population in North America has decreased from 1,354 birds per square mile in 2008 to 361 today. And the WSJ thinks all of us can adapt to these changes. Talk about the canary in the coal mine! Now what we must do to heed these warnings is to forestall our fear and move expeditiously toward the solution outlined in world plan. No more excuses! Just action.

I have taken you on this excursion because of its importance to the democratic (freedom) system. The U.S. Constitution addresses the problem that democracy must have an educated public and freedom of speech. But the writers didn't fully address how to keep oligarchs from financing falsehoods under the guise of free speech.

Now back to the NYT news regarding the Intergovernmental Panel on Climate Change which "found that decades of foot-dragging by political leaders have propelled humanity into a critical situation with greenhouse emissions rising faster than ever, though it remains technically possible to keep planetary warming to a tolerable level, only with an intensive push over the next 15 years to bring emissions under control, can we achieve this goal."

"We cannot afford to lose another decade," said Ottmar Edenhofer, a German economist and cochairman of the committee that also wrote in the report: "If we lose another decade, it becomes extremely costly to achieve climate stabilization" because CO_2 levels will be approaching 800 ppm.

The NYT goes on to report on the good news as well regarding aggressive action is being taken in many parts of the world, because it's becoming more affordable due to:
1. Energy conservation measures
2. Power cost for renewables is falling fast
3. The worldwide public is beginning to realize its importance.

Now back to the matter. These very strong measures can only be achieved with nuclear, with the caveat to achieve cheap 2C power.

Just one last review of a NYT, 4/22/14, article regarding energy, actually a full section of the paper, some eight pages. The only new information I gleaned from the paper was:

1. The average price for 1 kWh of power in the United States is 11 cents. The cost we discuss in Chapter 11 are for electricity generation cost only, and because the average price is close to the price for FF power, we can reasonably state that going from the generation cost to the average price, the difference (11 cents-4 cents = 6 cents or 80% of power) is generated from FF. Therefore, once we get to 80% nuclear, the average price will be about nine cents, instead of the 11 cents.
2. The article touts the benefits of renewables, mostly wind and a power regarding anticlimate gas reductions.
3. The Keystone pipeline will ultimately add more anticlimate gases to the atmosphere.
4. No discussion on price comparisons regarding FF renewables vis-a-vis nuclear. But, we already know the price in Germany is 14 cents.
5. Touting how the gas and oil supply increases will help the U.S. become energy dependent soon. Note: The United States has been importing 40% for 50 years.
6. I read and reread the article looking for some commentary on nuclear; answer: none. An eight-page section on energy with no mention of the word "nuclear." This proves our sequitur regarding the oligarchs have selected and purchased the information allowed for print. Irrespective of my criticism, it was a good article.

Related to our conclusions and difficulties getting truthful information, I was not aware that 15 states prohibit false political statements in campaigns (only 15 of 50). Of course, we agree that this should be a federal law and should be extended to the media and political activists (including EVs). But now opponents of these types of laws believe that these kinds of judgments are best left up to voters. So proponents like us call this "a right-to-lie position." Again, how can voters know how to vote if they are fed a diet of fiction and lies. We don't believe the first amendment means the right to lie; it is just like "shouting fire in a theater," which is against the law. In fact, we support legislation that would make it unlawful to lie with political speech, including activities.

So opponents have filed lawsuits against these "false" laws. A lawsuit filed against the Ohio Law will be heard by Supreme Court shortly. These laws should be considered "self-evident." However, we already know the opinion of Justice Alito regarding any attempt by the government to penalize purportedly false speech in political contexts would present a grave and unacceptable danger of suppressing truthful *speech (United States vs. Alvarez)*. If the courts rule that these laws violate the Constitution, then the Constitution needs to be amended. Related to this matter, how can the public be expected to vote on nuclear power if the oligarchs are allowed to spread falsehoods/fear. Democracy is built on an educated public making these decisions. Otherwise (?).

Another person that opposes these laws is Rush Limbaugh. He doesn't want anyone challenging the falsehoods he expresses on many days. Just like him calling Sandra Fluke a slut.

Recent court rulings have made it easier for the media to authenticate the falsehoods, just like the false news that is reported every day by Fox News. And they try to cover it up by reporting to the public that they are "fair and balanced" when everyone in the world knows they report only what their owners believe in. I will give you only an example on false reporting "without getting into politics." After the airplane crash in Reno, Nevada, I was very interested because I am a pilot, and I wanted to hear what happened.

1. Fox News reported that twelve people were killed and many were injured.
2. CNN reported that three were killed and seventy-five injured.

So I went to sleep, wondering what the truth was. I woke up the next morning and read in the paper that three were killed. I turned to Fox News. This channel never reported anymore on the matter after that.

Our revolution regarding the merging of the best features from the Western democracies, and the best from the Guardian States may be the only way to solve the problems with Western democracy constructions before the Guardian States gain more power. The recent Supreme Court rulings regarding allowing oligarchs to finance our political system and now with the court hearing the lawsuit to overturn state laws that criminalize the publication or advertisement of false information, this may be the final straw to start the revolution. If it's true, then our Constitution guides decisions like the one above or the decision in 1878 that would not permit blacks the right to vote because they were considered the property of oligarchs. So we already know our Constitution has outlived its usefulness, especially if the court rules that false information is allowed under Article 1 (free speech).

Today, 4/15/15 (last entry, I think)
NYT

Reports on the West drought in California, Arizona, Colorado, and New Mexico. Water levels in reservoirs are at 3% of capacity (some). And the temperature has risen twice the global average. And expected to rise another 5C (10F). Worst drought in 13,000 years.
WSJ - Opinion

Complaining about Obama's carbon reduction guidelines regarding too much cost and should leave it up to the states. (Mr. Baxter, CEO, Ameren Corp.)

They make slaves of all of us. Some refer to the Supreme Court as a group of politicians with robes.

So I will sign off for now on this chapter. If I write another book, I will use the knowledge I gained from with this research and title the book as "The New Modisized Western Democracy—The Bottom-Up Approach." The mission will be "to integrate the knowledge, wisdom, and power that we have available to the world today into a new modernized Western democracy for the combined, integrated. world collective to be guided by."

20.0

Nuclear Power Safety Issues

Now we turn to the primary reason why the public is against nuclear power. *Fear* and we all know what fear can do to a person's cognitive competency. EV take advantage of these feelings with lies; nuclear has already killed millions.

As stated elsewhere herein, there's still technology and empirical knowledge to be developed so that nuclear power is 100% safe. But these are solvable issues. And these solutions can be built into regulations. For example, Fukushima would never happen in the United States. Why? Because we have regulations that require engineers of NPP to design for earthquake and tsunami issues. That's not to say there's not a lot of knowledge that has been developed since the accident in Japan. Engineers and scientists will incorporate this oeuvre into future plants and improvements to the others. Regarding Fukushima, over 20,000 people were killed because of the tsunami. Only six workers at the NPP were exposed to high levels of radioactive water. And none have died yet.

The NYT published an article by David Ropeik on October 22, 2013, p. A21, titled "Taming Radiation Fears." Mr. Ropeik is an instructor at the Harvard Extension School, and author of the book *"How Risky Is It Really? Why our fears don't always match the facts."*

Mr. Ropeik has studied both the Fukushima and Chernobyl accidents, and he reported on the misinformation regarding Fukushima and the fearful reactions that have spread around the world.

1. Scientists report that the radiation from Fukushima has been relatively harmless, and none of the six workers were killed or developed life-threatening cancer. This is consistent with scientific studies that the radiation from these accidents do not cause excessive danger to workers or communities. Scientists have studied the entire population exposed to atomic bombs; some were exposed to extremely high radiation levels. Only 527 deaths out of a population of 100,000. That's less than 1%,

much less than exposures from many harmful substances in our environment.
2. The survivors from Hiroshima and Nagasaki have shown us that ionizing radiation,
(i.e., the type from nuclear reactions) is not a type or radiation that causes life- threatening cancer or genetic mutations.
3. The most important aspect of the research on survivors is the conclusion that the dose levels below 100 millisievert causes no detectable elevations of cancer or deaths.

The Chernobyl accident demonstrated for us that "the population within a six-mile radius were exposed to radiation levels less than 100 millisievert, causing no detectable levels of illness or cancer. And Chernobyl killed only a few, not the thousands and millions the media and EV write and protest about."

These results from the worst of all nuclear power plant accidents should be the evidences we need that nuclear power plants are "far and away" the safest form of power that we have. Read Mr. Ropeik's book for a complete view of how fear keeps us from making good decisions. Future NPP will include the lessons learned from these accidents. Future NPP will be even safer. After all, this is what our Lord gave us (i.e., the evolution process to make life easier) the longer we work on problems.

Another good research on radiation is Mr. Nelson's book, *The Age of Radiance: The Epic Rise and Dramatic Fall of the Atomic Era*. Mr. Nelson researched the relationship between radiation cancer and deaths. He concludes that radiation risks are less than 1%, consistent with Mr. Ropeik's results. And radiation risk, even to nuclear operators, is less than the risk of being a real estate agent or a stockbroker.

The lady that defined radioactivity, Marie Curie, wrote in her thesis on radiation: "Now is the time to understand more, and fear less." For our part, we say: "Now is the time to expose the lies/fear of the EVs, and to collectively vote out Big Oil and plutocrats." This includes both Eastern and Western Countries. And once the nuclear renaissance is underway, the people will have more power than King Coal or Big Oil. See how all this fits together? Through the combination of power from knowledge and nuclear, the world collective will have the energy and passion to eliminate every monarchy or oligarchy wherever it exists. And we will release enough radiation to eradicate all of our fears.

The best way to express the relation between fear and action is to review the poker game analogy discussed below. This research on fear is what led us to the relationship between knowledge, creativity, and wisdom. With a lot of research and hard work, all the solutions to these difficult subjects start to come together.

More can be learned about our philosophy by reading about the fear factor related to why nuclear power is not accepted in some parts of the world, and how our research led us to a better understanding of the relationship between fear, action, knowledge, creativity, and wisdom.

The best way to express these relationships is with a good "skillful" poker game.   The best choice to make is sometimes to choose the risk path over the safe path. In a poker game, if you cannot "go all in" when the situation demands it, you will never be a long-term winner. Once your competition identifies you as being "afraid" to place the proper bet, you will be killed immediately.

It's just like coal miners; they place their bet every day, knowing that the safe-path is to stay at home and join the 47 percenters. If the miners stayed home, none of us would have the lifestyle we enjoy today. This shows us that knowledge and experiences lead us down the "wisdom path" faster than an elite education studying the Ayn Rand philosophy. "We already know the Ayn Rand education is for self-serving reasons rather than to obtain "true knowledge."

Before you can hope to have the knowledge to play the game, it's been said you will need to play 60,000 poker hands. At a minimum of $10 per hand, your investment will be $600,000. Easy math (without a calculator) to obtain this knowledge. No formal education is needed, just a little heuristic learning.

Even at this experimental level, you are more than likely going to lose many games before you win one.

And fear comes from playing the game without knowledge. Without knowledge, we are all losers.

This fear will come from the "lack of knowledge." So if you don't have the funding to gain this knowledge, don't expect to be a winner. Just play for entertainment (with money) you will *never* need.

The poker rules and procedures can be learned quickly in a school. Knowledge will only be gained by the experience of the 60,000 hands. So it's the experience that matters in life.

This subject of science and knowledge has been debated since the scientific revolution. A book on the subject, *The Lagoon: How Aristotle Invented Science* by Leroi makes its case for Aristotle. At the same time, he describes the history behind how the early scientists. For example, Copernicus, Kepler, and Galileo. had to work extra hard to overcome the "sense perceptions" presented by Aristotle without any data or numbers to confuse the science as he sees. Leroi, being a biologist like Aristotle and with its biases, states that "Plato's science is barely distinguishable from theology." This is so with one being enamored with the beauty of math and the other with the beauty of sensed perceptions. This debate has only intensified with all the knowledge we have today. And once this debate gets mixed in with politics, it's hard to recognize where the knowledge is.

What about creativity? Creative people organize their lives around repetitive, disciplined routines. Order and discipline are the prerequisites for creativity and daring (no fear).

Now when it comes to wisdom, that's covered later. As you can see, this research is leading us to all kinds of knowledge.

Other EIA information

The last Copenhagen conference on climate change in 2009 fell apart because big developing countries could not agree to accept legally binding commitments. They did, however, pledge country-by-country targets.

The United States is responsible for 25% of the carbon emitted (to date) China is second with 10% but closing in fast. The country pledged a 17% reduction by 2020. This was crucial in order to get pledges. It is halfway there.

Japan is falling behind because it shut down its nuclear facilities after Fukushima. China made no pledge on carbon emissions.

By 2040, CO_2 emissions are projected to grow by
USA - 0%
China - 2.1%
India - 2.3%

Japan - 0.1%
Germany - 2.5%
Total non-OECD - +1.9%
34 members of the organization for Economic Cooperation and Development (OECD)

Germany's CO_2 is increasing because it closed its nuclear plants after Fukushima.

The WSJ came as close as it could to endorsing nuclear by claiming on 7/2/14 "that FF energy will be outmoded over some period by cheaper alternatives. To boot, the world will discover that climate change is not the challenge facing it after all." And the WSJ knows conventional alternatives are more expensive. So what's left for this WSJ opinion: *nuclear energy*. We are close to having the WSJ in our collective, but we will have to keep it secret.

Obama - New EPA Rules, 6/2/14
30% reduction in CO_2 gases by 2030

Paul Krugman writes: "This will reduce the economy by $50 billion. This will affect the 600 U.S. coal power plants the most." Also, this is a compromise with the desires of the coal industry and the EAs. The rule details will be announced on Monday (6/3/14). Both the NYT and the WSJ reported on the rule's impact and concerns regarding (1) politicians in coal states, (2) the health and environmental benefits, (3) the economy, and (4) China and the United States. The U.S. Congress is taking efforts to block the rules. Neither newspapers discussed nuclear power. No one but you and I are fearless enough to discuss nuclear as the only path available. I believe the power industries will support nuclear as long as the public and politicians will grant permits. This will require all of us to confront the fears and lies of the EVs. The power industries understand the economics of nuclear; the engineering/construction companies are ready to build the new (better) nuclear power plants.

In the same WSJ issue, also discussed was U.S. manufacturing's optimism about the future. One of the reasons is that energy cost in the United States is 50% less than Germany.

21.0

Philosophy of Life

21.1

Ayn Rand

I added this section to the book because I kept hearing many politicians quoting Ayn Rand as their guiding philosophy for politics, especially on antiscience (climate), etc., so, I wanted to know more about her philosophy. I read three of her books. The Ayn Rand philosophy in one simple statement that boils down to this: procapitalism, pro-individual, probusiness, pro-oligarchy, anti-altruism, anticollectivism, antireligion, antigovernment, and antidemocracy. Since the first two antis are so important, here's the definition for these:
Altruism

1. Unselfish regard for or devotion to the welfare of others
2. Behavior by an animal that is not beneficial to or may be harmful but that benefits others of its species

Collective

1. Denoting a number of persons or things considered as one group or whole
    a. Formed by collecting (aggregated)
    b. Of a fruit (multiple), maybe gays also.
    c. Related to a group of people (individuals)
    d. Involving all members of a group as distinct from the individual's collective action.
2. Marked by similarity or with the members of a group
3. Collectivized or characterized by all members of the group

---

21.2

Johnroy Messick: The Renaissance man

After 60 years of working like a drone, my philosophy can be boiled down to this:

This world has some 10 billion people (2 billion without electricity). We are all brothers and sisters. The United States is great because we have a government that is a mixture of capitalism and socialism. Also, our Constitution mandates the government be responsible for the general welfare of its citizens. We all have a duty, like the Bible states, "to care for the weak and vulnerable." I extend this to the world; other aspects of my philosophy can be surmised from writings herein. You will observe I'm more like Aristotle than his critics, i.e., another Peripatetic roaming the world, gathering knowledge wherever it resides, then sharing it via heuristic learnings.

Ayn has a continuing and renewing following by virtue of rich parents sending their offspring to the same elite schools the parents went to. These elite schools all teach the Ayn Rand philosophy: learning how to rig the oligarchic environment so that the rich don't have to pay their fair share of taxes or be swayed into giving any of their wealth away. So this becomes a continuing way for the rich to grow their wealth, being financially able to send their offspring to elite schools that offer studies in the Ayn philosophy. These Ayn cult members never expose what the Ayn philosophy really is. Otherwise, they would never have a following. That way, the wealthy have a means to identify each other just by saying they believe in her philosophy.

Look what happened to Romney when he was delivering a speech only to members of the Ayn Cult. Jimmy Carter's grandson recorded Romney as saying 47% of the people are "takers" and would never vote for him.

Romney is a good example with the 47% comment while at the same time getting richer by buying U.S. companies and selling the companies to the highest bidder; China could pay more because China could move the plant, technologies, and jobs via lower-paid Chinese workers.

Paul Ryan recently stated, "I feel like we are living in an Ayn novel," indicating that the rich have to give some of their wealth to the undeserving lazy people of the world. Ted Cruz also acts like an Ayn Rand cult member.

No wonder Romney chose Rand Paul as his running mate. And he thought they could win elections in districts that were not gerrymandered. Better go back to school on that one!

And we don't have that many scientist/engineers. Today, as indicated in the WSJ, 11/11/13: We graduate about 76,000 per year. This is one person per 200,000, as estimated using data from the Department of Education. The problem Ayn Rand has is that engineers are a collective of working drones. Ayn thinks of us as the new intellectuals. She knows little about the working population of the world.

I like to use a beehive analogy for my philosophy. I know a little about bees. I learned this when I was a young man, and I had a friend in Appalachia. His family had eked out a living by operating a small sawmill to make lumber to build pallets and to construct beehives. They would build the hive then harvest the queen bee from the mountains. They ran a transport delivery business to deliver the hives to customers in Florida. No one would deliver for them because the roads were too muddy. The family—from grandparents to babies—lived in their own shack built from the sawmill lumbers. They owned a couple of worthless acres on a hillside.

Back to the bees, the family decided they needed to learn more about bees. So the patriarch built a beehive and installed the hive on a window lattice in the living room, like a palace evermore with the window pane in front of the hive. The entire hive could be seen. When I visited him, we would sit for hours observing the activities of the hive civilization. Yes, I said "civilization" because the bees worked as a collective. Every bee had its place in the hive. Some were builders; some were spotters (i.e., the bees that search out new locations for the new queen bee). Some would carry off the dead to their final resting places, some guarded the entrances to protect from invaders, and some repaired the hive.

The talented bees would engineer the hive location and build around any suitable location like a lattice. The loving bees would care for the sick and wounded. All the bees had a job, and they all got paid the same. Some bees would quit working and would be lazy around the hive. The lazy bees would still be allowed to enjoy the hive and suck on the honey. I examined one of the lazy ones and found that the bee's wing was injured due to all the flying. We all could learn from the bees. Ayn's philosophy has no place for the bees.

So, just like the bees, other living species can survive; they just can't invent. That's why human scientists and engineers are so important to society. By having a political system that disregards the recommendations of scientists and experts, we might as well be bees. And now I know why Ayn Rand doesn't like collectives. The main reason is this: If the hive owner (business person) doesn't take care of the population, the collective will sting like hell!

The situation actually occurs with both political parties in the United States.

Republicans are doing this bit on climate change. Democrats are doing this with nuclear power.

And now we read in the NYT, 12/2/13, reported that climate change is killing the bees! This may be more serious than the concerns on oceans and storms.

Now that we have explored knowledge and creativity, let's take a peek at wisdom. It's related to the matter herein: Vivian Clayton a geriatric neuropsychologist scoured ancient texts and found that people who are considered wise have three characteristics: cognition, compassion, and reflection. And as people age, they have more information in their brain. As is explained in the book *The Wisdom Paradox*, cognition can utilize this information and develop pattern recognition that can be used to form the basis for wise behavior and decisions. This tells us that inviolate and true wisdom can only come from people who have lived long enough to have this depth of knowledge in their minds (reflective).

What about compassion? We can use this knowledge to understand and to help others. Dr. Ardelt, a professor of sociology at the University of Florida, expanded the research on wisdom with older people to explore those that showed wisdom. She discovered that the main impediment to exploring this wisdom was coping skills. The reason is that as we age, many wretched things can happen. This impediment can lead one to think: "I'm not who I used to be, so how can I be useful to myself or others." She goes on to say, "Wise people try to understand situations from multiple perspectives, not just their own, and they show tolerance as a result. You're not focused so much on what you need or deserve but what one can contribute."

Professor Laura L. Carstensen, a psychology professor and founding director of the Stanford Center on longevity in California, states: People that rank high in neuroticism are unlikely to be wise. They see things in a negative and self-centered way. This leads to not being able to benefit from the knowledge they have gathered." To this, we add: "Knowledge needs to be shared so that the reflection aspect of knowledge can grow."

Daniel Goleman, author of *Focus* and *Emotional Intelligence* says one aspect of wisdom is "having a very wide horizon that doesn't focus on ourselves." An important sign of wisdom is "generativity," a term created by the psychologist Erik Erison. He developed an influential theory on stages of the human life span. "Generativity" means giving back without needing anything in return. "The wisest people do that in a way that doesn't see their lifetime as limiting when this happens" (NYT 3/13/14)

To continue, when Erikson reached old age, he found it necessary to add a final stage of development to his understanding of life—one that "depicts an old age in which one has enough conviction in one's own completeness to ward off the despair that can come from old-age gradual physical disintegration. This can be accomplished by relying on the knowledge gathered." This "wisdom" knowledge leads us to this conclusion!

Those who believe and live the Ayn Rand "philosophy of life" can never achieve wisdom. Her philosophy is based on the individual being the center of the universe, a violation of the first tenet of wisdom, and the follow-up that this leads to is no compassion for anyone that doesn't add value to the wealth of that individual (the third tenet of wisdom). So to answer my question: Why do right-wing libertarians disagree with experts in a particular field? The answer is "they cannot have wisdom because they have been trained in the wrong philosophy of life." If I had not researched Ayn Rand, I would not have stumbled onto this sequitur.

So now we have all the ingredients for a successful renaissance, and at the same time, we understand that this is a path to greatly benefit the world. We will all be working on this renaissance without any "monetary" returns, just personal compassion for the ailing, without electricity.

When I started my research about the "Philosophy of Life," I had no idea what I would find and where it would lead. Now, after integrating the science of wisdom, the light came on: the subject of experience and knowledge then leads to compassion with passion. So if one has not reached the level of compassion with passion, then one has not reached wisdom; one may be very intelligent, but it will not be recognized as wisdom. Of what use is knowledge and experience if not used to help the real problems of our time. It's not only unrecognized; the wisdom is not there. The books referenced herein provide a full description. This work on wisdom has added to our knowledge and experience as well. Even this introduction adds immensely to one's contemplation of the value of life.

So let's all extend ourselves to the level of compassion with passion. Our contribution to the subject of wisdom, is adding "with passion" to compassion.

22.

Economics

This chapter was in the index from the start of my research. Economics is not only for economists and politicians but also for the public and scientists/engineers. Also, I started this chapter before the young French economists, Piketty, Stefanie, and Saez (henceforth referred to as Piketty) came up with their research using the science of "big data." Big data refers to using the capabilities of search engines to gather mounds of data and then using computers with algorithms to crunch the data. Economists then analyze results. These tools were not available prior to the IT revolution. So now, with this knowledge, I almost have to start over with our economic discussions.

Piketty, in his book *Capital in the 21st Century*, uses big data to settle the contradictions (economic theories) such as the historical theories between Karl Marx and Adam Smith and the more recent adaptation of these theories between Milton Friedman and Maynard Keynes. With this knowledge, economics can be considered a true science without contradictory theorems. It will be up to the economic scientists associations to support the theories as "the science" for economics. This will not be easy; many older economists will have to abandon what they were taught as students of economics, and many oligarchs have invested heavily in theorems that have been proven to be incorrect.

This is very important to our matter. Piketty defines a methodological risk path forward that could very well save democracy from the weaknesses pointed out herein.

Piketty with his work asserts on how wealth must be managed by big governments, and now with this research, affirms energy must be managed by big governments. The combination of these scientific principles can bring the world out of poverty "today," which is the ultimate goal of democracy. This is not a utopia but a place society can reach with a trammeled democratic capitalistic society. This is what Piketty believes, and what we believe also. Once this inequality issue is settled, nuclear can provide all the energy our collective needs to continue advancing forward.

I call this the fifth dimension of our book, after the quadrivium of cheap energy, media, politics, and environment (the 5Es).

This inequality issue started with me in high school. I worked two paper routes and a job as a grocery bagger. I didn't have much time to study. Then in college, I became a co-op student with St. Regis Paper. But I still had to work as a grocery store bagger in order to pay 100% of my college expenses. Again I had to work instead of studying. I am jealous of those that didn't have to work. I'm not saying the work was harmful; it's just that I didn't have the time to learn the philosophies that bosses wanted from their educated employees. Plus, I could have been a better writer earlier in my career. Maybe that's why I enjoy continuing to develop and extend my knowledge for the benefit of future generations.

Another story that fits. I was working for Wynn Dixie Grocery, the week before the minimum wage was implemented in 1959. I worked all week for 50 cents/hr. I worked 40 hours and was paid $18.02 for the week. The next week, the minimum wage rate was increased to $1.00/hr. My paycheck was $36.50. This helped me immensely while in high school. And I didn't lose my job. So how did I become a raging progressive liberal today? By experience, not by a sophisticated philosophy. This is the value of being a solipsistic Hegelian, just to learn the words to write good fiction (e.g., Mr. Ridley).

Another reason for bringing this economic science forward is because I want my readers to extend their portfolio to this knowledge. Even in some of the media outlets, this knowledge is suppressed. Why? Because the oligarchs are on the wrong side of the proven theorems (e.g., taxes to help with the inequalities). Examples show up in three major papers: NYT, WSJ and LAT. This fits in so well with our previous discussions regarding the media's reporting on the IPCC report on climate change.

First the NYT. Paul Krugman has written reviews on Piketty's book. He wrote that Piketty takes on Marx and Smith with Piketty's conclusion that "untrammeled capitalism" in the West provides more wealth to the owners than the workers. The money does not trickle down, and Piketty proves this using big data analysis. Paul didn't go as far as I had (e.g., bringing in economics into the scientific community).

Now onto the WSJ. Just like with the IPCC report on climate change, the WSJ has been silent on the Piketty economic news. Then the journal got an antiscience, antigovernment fiction political writer from Britain, Mr. Ridley, to write some fiction on the matter.

But Paul did state that the Piketty book is "more than just economic information" and is "revolutionary." It's amazing how evolutions work. Just work hard and keep at it. Then sit back and let it evolve.

One last comment here. Once these amazing tools are available to all and once we have the wealth to educate all to a good level, our collective will have the power to grow. We know cultural wealth is more important than material wealth.

In support of Mr. Piketty, Krugman (NYT, 1/30/14) wrote about the *wrong* predictions of top economists, like Art Laffer and Steven Moore (both work for the Heritage Foundation and ALEX):

1. Five years ago, "we expect rapidly rising prices and much higher interest rates over the next four or five years." Look at what actually happened.
2. Recently, advising Kansas to slash income taxes to jumpstart an economic boom, they roared, "Look out Texas!" Instead, they got a Moody's downgrade.
3. Historically predicted, Clinton's tax increases on the rich would cause a depression; instead, we got a spectacular economic expansion.
4. We have read many of their articles in the WSJ, but none "retracting how wrong they were." How knowledgeable does the public need to be in order to unravel the misinformation and discover the truth.

After stumbling onto this economic knowledge, I now know why I was fired from P&G. I made a presentation about my project, "Densified Detergents," and I made the statement that "profits are not the only thing to consider." And then, I marked-up a book that was mandatory reading for managers. I marked-up the book because the writings were all John Burch Stuff, and I disagreed vehemently with the statements. The next week, I took a psychological test. My manager called me into his office and told me, "See the lines on this graph, yours go this way, and successful managers go this way!" Three months later, I was fired. So after digging my way out of poverty, I was expected to have the same philosophy as the elite managers that were studying philosophy instead of working. Another way the oligarch's wealth comes through. I never made that mistake again. "Never expose one's politics in the workplace," even when I worked for a Jewish owner/president that was kicked out of Cuba during the revolution. I later became the youngest partner in the engineering/construction firm.

Economics are for collectives also. And once we understand the matter in Piketty's work, we can proceed with solving our democratic weaknesses. Communism will be set back again. If we don't, capitalism/freedom will lose.

After discovering these weaknesses with Western democracies, I did some more research on this matter. I read a book published by Freedom House, a human rights group, and written by John Micklethwait and Adrian Wooldridge, titled *The Fourth Revolution*. They wrote that "recent" developments exposed Western weaknesses, citing recent Supreme Court rulings, inactive Congress, and low taxation (e.g., not paying for Western values or principles). This started after the fall of the Soviet Union. Western democracies went into an era of complacency. At the same time, the Guardian States: China, Russia, etc. have become "modernized autocracies, gaining ground on Western democracies related to:

1. Long-term planning
2. Speed at the top, without the need to get public agreements
3. Relying on experts to expedite programs

This has led to these guardian states to outpace Western democracies via innovations and big programs, at the same time realizing there are flaws with a modernized autocracy, as follows:

1. Lack of freedom
2. Corruption
3. Speed of action at the bottom

That's why the authors suggested that a fourth revolution is needed to integrate the two systems into one that has the best features of both and at the same time, eliminating the flaws inherent in both systems.

The democratic flaws are the following:

1. No long-term planning

2. Not willing to pay for government services (e.g., higher taxes for desirable services)
3. Education needs

And these thoughts regarding experts and governing were proliferated by Lippmann, an ex-aide to President Wilson and a prolific writer, prodigy, and friend of Maynard Keynes. He later criticized LBJ for trying to finance the war on poverty and Vietnam, including guns and butter. The collapse of the gold and dollar-based international monetary system and the breakout of inflation after World War II confirmed his wisdom.

In addition, he was a close friend with Lewis Douglas, a staunch defender of capitalism. He applauded the amazing feat of capitalism and big companies converting from wartime to peacetime operations. He advocated:

1. Ameliorating downswings in business cycles with deficit spending
2. Free market ideas espoused by Friedrich Hayek

All of this led to Lippmann believing that:

1. The press (media) is a big part of the problem (e.g., sloppy coverage of government policies and actions).
2. Better to have top-down rule via experts, bypassing the need to educate the public.

All of these are covered in Mr. Goodwin's new book, *Walter Lippmann: Public Economist.* One more nail in the coffin.

Supreme Court

And now we have others supporting our thoughts on the Supreme Court, with the book *The Case against the Supreme Court* by Erwin Chemerinsky. He described "in detail" how the court has failed "to enforce the constitution against the will of the majority; and thereby do good for the country." We believe a practical solution is to have term limits so that the judges become more up to date on the progression of the country. How can 75% of the judges over 75 years old be up to date on a country that has only 10% of its population aged over 75? And they have jobs for life, not to mention how judges can live forever with the easy life they have, totally insulated from the inequality of the country. More on this subject later.

This knowledge has led us to one of our conclusions that at an *evolution revolution* may be needed. And we have heard others smarter than us describe the judges "as nothing more than politicians with robes."

Without age limits for judges, the natural evolution process is stopped dead in its track. So when the revolution happens, the Supreme Court will be one of the first stops along the way. The writers of our Constitution came before the teachings of Darwin. So this evolving knowledge needs to be considered in our new constitution.

Here's a most recent example of why the death of Eric Garner was a work by the police. The Supreme Court recently authorized Mr. Garner's death when it overturned "a no-chokehold" ruling in a 1983, *City of Los Angeles v. Lyons*. The lower court in Los Angeles ordered "the police not to use chokeholds unless an officer is threatened with death or serious injury and to institute better training and record keeping." It's just another example of the court not evolving with time. Congress could do away with the job-for-life provisions, but our do-nothing Congress is too busy suing Obama.

And just this week 12-23-14 NYT:

The incompetence of the court was pointed out by a Supreme Court judge. Scalia was found incompetent with rulings that were inconsistent. Scalia admitted his incompetence, but then went on to claim his incompetence will not affect future rulings. Without the problem we have with our constitution, Scalia would be fired and sent to the Woodshad. Maybe we should start the revolution sooner than I thought. Otherwise, final rulings will continue to be made by an incompetent judge. Without a revolution, the United States will be guided by incompetence until Scalia dies. Look at what happens when man interrupts God's evolution process: this causes incompetence rather than the natural selection process that weeds-out incompetence. If not for his incompetence, the United States would <u>not</u> have been involved with two wars, <u>and</u> the anticlimate emissions would have been solved two years ago. And the WSJ would have never reported on this matter.

And then, after writing about Piketty, the NYT, 4/25/14 had two editorials, one by David Brooks, and the second by Paul Krugman. Both discussed how the Piketty phenomenon is more than a good book; it's about truly bringing economics into the scientific community (my words). Piketty has used big data to support the economic theories that work. And as Paul states, "And conservatives are terrified." The reason is that some of Marx's theories about capitalism have been proven (our words again). Also, these theories support Keynesian economics, which William Buckley blocked from teaching not because it was incorrect but "denouncing it as "collectivist." And now Piketty has proven that the facts are on the side of "the people."

The WSJ has reported on Piketty but only to criticizes his work as "segueing from Mr. Piketty's call for progressive taxation as a way to limit the concentration of wealth—a remedy as American as apple pie, once advocated not just by leading economist but by mainstream politicians, up to and including Teddy Roosevelt—to the evils of Stalinism."

How can I ever finish my research on this matter? At least now I understand why Ayn hates collectivists! If I had a boss, I'm sure he would fire me. But keep in mind that none of the historical economists had the tools to collect and analyze big data. Will Piketty be credited with saving capitalism? And for those of us on this journey, this is so important because we don't want the Guardian States to dominate the world with nuclear power while the West is curled up in a ball, afraid to step on the risk path. Now let's see if the collective can get smart enough to accept the new science of economics. I don't have much hope unless we move forward with our world plan, especially, the two gems included in our conclusions/recommendations.

Our Work on Economics Afore Piketty

The U.S. economy today (2013) is in the midst of an awful downturn, and according to Paul Krugman, we are still in the second Depression, with unemployment and inequality at all-time highs. And I can't help believing that the amount of money we have sent to the Middle East is a big factor, both directly and indirectly. These dollars are used by countries like Iraq, Afghanistan, and Iran to sponsor terrorism and other warlike initiatives. And our response is to double-down with wars. The direct effect is in all the monies spent on imported oil over the last 50 years. Just in 2008, we spent $500 billion, down to only $327 billion in 2013, with a projection of $250 billion in 2020. The indirect effect is the money spent to fight these wars.

So with the United States spending directly about $4 trillion in the last 10 years, this money could have stayed in our economy. With a turnover multiplier of 4, our economy would have about $16 trillion more, almost as much as our national debt.

And now, I have to criticize Paul Krugman regarding his NYT article on 4/18/14, "Salvation gets cheap," where he wrote about the IPCC report. He wrote about the positives and how the world doesn't have to sacrifice to reduce anticlimate gases because renewables are getting cheaper and can replace FF. With no mention of nuclear, what do we expect? The NYT can't go against all the EV of the world. The newspaper is fearful of losing a big part of its constituencies. Advise to Paul: Stay with economics and don't venture to far into energy, and I promise to do the same with economics. Although I must admit, Paul was just writing about the IPCC report, and the report does say that renewables can eventually replace FF without any mention of nuclear either. We have to realize that this report is written by climate scientists not energy scientists. Paul could spend more time on his colleagues' economics work.

Regarding one economist honored for work on media slant, Mr. Matthew Gentzkow, he was awarded the John Bates Clark medal, one of the most prestigious awards for economists under the age of 40. Mr. Clark wrote about younger economists using big data to form sequiturs, not just the theoretical ones, like Keynesian or Friedman theories. So, Paul, don't get left behind!

Regarding the WSJ, it apparently reported on April 7 on "second climate change." I missed this one, but Jonathan Lynn from the IPCC wrote a letter to the WSJ, refuting the second climate change article wherein the IPCC was said "sexing up" the summary for policy makers. Mr. Lynn informed the WSJ that the IPCC has a mandate not an agenda. He then told the WSJ of its many inaccuracies, including statements that were not a part of IPCC report. The main point here is this: Because of the WSJ's previous editorials on April 7 and the criticisms it brought, the paper has not reported on the IPCC report released on April 14. And now we know why.

China is already seeing measureable productively gains because of its electric bullet trains reducing emissions and providing better economics because the trains are powered with electricity instead of gas or diesel. In the future, China's power plants will be mostly nuclear. If China and India accomplish their goals with nuclear energy and the United States doesn't move toward more nuclear energy, it will be no better than a third or fourth economic power. If Russia or Brazil get aggressive with nuclear power, the United States could be no better than sixth. This not only means a sixth-rate economy, but will also cause high unemployment due to our exports being too expensive to manufacture (written before research on Russia, see Chapter 14).

What's really exciting are the prospects for the future. The United States will overtake Russia as the largest oil producer in 2014, if we can make progress with nuclear (move forward on the risk path). We could be the powerhouse for the world. Notice how I said, "for" instead of "of." It's because of these advancements in oil from shale and fracking for natural gas that we find ourselves in this position. Engineers and experts developed the knowledge for these advancements. Since 1974, energy efficiency led by the United States has saved $1.5 trillion in FF purchases. If we had used the money to build new NPP, the United States would already be there, and all these discussions about the expense to reduce our carbon footprint would be over. The same could be said for other big polluting countries of the world. Of the top three polluting countries, the United States is the only country that is not on the 2C risk path. Shame on us. All because of EVs protesting with lies and fear-mongering.

More on pre-Piketty economics. The 2010 models were built to explain unemployment and the job market, but these didn't solve how to reduce unemployment, just an explanation of how it can happen, etc. to buyers and sellers of jobs. Two Americans and one British, Dale Mortensen, Peter Diamond, and the 70-year-old appointed to the Fed by Obama were behind these models. Their objective was to make it harder to hire and fire (i.e., sclerotic labor). Like a lot of economic modelers, they assumed a frictionless process. In Europe, this is especially not true. It's far too difficult for employees to move to where the jobs are.

How did economists get it so wrong? Just before the Great Recession, economists were sure they had solved the macroeconomics problem (i.e., recessions/depressions). In 2008, Olivia Blanchard of MIT, now the chief economist for the International Monetary Fund, declared that the state of macroeconomics is good. Even Ben Bernanke, a former Princeton professor, celebrated the moderation by pointing out the economic performances over the last two decades prior to 2008. This contributed, in part, to improved economic policymaking.

Then in 2008–2009, everything came apart. Not seeing the catastrophic problems coming was not the real problem.

So why go any further! The lessons of Adam Smith and Maynard Keynes were lost in the fancy algorithms. Smith, in his book in 1776, *The Wealth of Nations*, has models which were followed by the Great Depression. Keynes wrote on how to correct Smith's 1936 book, *The General Theory of Interest and Money*. His intent was to fix capitalism

Another opinion in the WSJ, 6/23/14:

It's a rehash of Piketty's book, then the opinion starting with the statement:

"Abstract notion of inequality being too high." So with that proposition, one doesn't need to read the opinion!

Joseph Stiglitz, another Nobel Prize winning economist, sums up his work on his book *The Great Divide* in two chapters, "Inequality Is Not Inevitable" and "Inequality Is a Choice." When I first saw Mr. Kristof's article in the NYT dated May 3, 2015, titled "Inequality is a choice," I thought it was a typical piece about choices impoverished people make that keep them from climbing out of poverty. But Mr. Kristof was covering Mr. Stiglitz book and goes on to point out Harvard's book *Inequality* that points out 15 steps that can be taken to reduce inequality. These steps are taken from Mr. Stiglitz's general philosophy that "Inequality is a matter not so much of capitalism in the

20th century as of democracy in the 20th century focusing on making inequality a priority and what policies to turn to." Our addition to this excellent discussion is

One, policies can't be changed until politicians or leaders are changed. Not the WSJ opinion that "Inequality is not rapid."

Two, integrate the value of the "people's power" into these inequality solutions.

NYT, 6/27/14

Paul Krugman wrote about information from Bill Kristol's quotations that show that the critics don't believe their own positions. To err is human and to persist is diabolical, as can be gleaned here:

"Clinton-care might be successful; damaging the small government arguments geared to tax cuts."

It's about politics and ideology, not analysis. The middle class can be defeated by restraining governments. And now for us, with the media lying to the public, make it very difficult to decipher the fictitious opinions.

More criticisms of Mr. Piketty's work:

The *Financial Times* (FT) 5/24/14 reported that some of Piketty's math "may not be correct." Chris Giles, the economics editor of the newspaper, wrote "the investigation undercuts Piketty's claim that wealth inequalities are headed toward those prior to World War I." Mr. Giles cited Piketty's thesis that "an increasing share of wealth is held by the richest few." Mr. Piketty responded to his "openness to second-guessing the newspaper's analysis was possible because he had posted his data online, and that subsequent research using other datasets had confirmed his results."

NYT, 5/24/14, p. B2

Piketty went on, challenging the FT to show different data disproving his conclusions. Piketty is confident enough about his research that he will change any conclusions if needed, with facts not subpar visions.

We know that when the oligarchs are faced with such evidences, they will turn the world upside down to at least obfuscate the issues so that laymen will not know which side is correct. This is just another example of what money can do to sway the electorate. It will be up to the economic-scientific community to sponsor the truthful forums. The oligarchs may pay-off the truth by paying the media and economic activist to spread lies and fear. I believe Piketty and crew have the courage to withstand the assault. If the oligarchs overtake the Supreme Court, the "evolution revolution" will start immediately.

More support for Piketty's economics:

Now, the Nobel Memorial Prize in Economic Sciences has been awarded to another French economist, Jean Tirole, for his decades of work proving that "markets are not perfect," disproving the pre-Great Recession science taught and founded at the University of Chicago (i.e., "markets are efficient"). He has proven via big data that because of the lack of regulations, banks and financial companies started the Great Recession. This is the second year that the Nobel committee has honored an economist from France. Piketty was the first. Both were honored for their work on imperfect markets and the need for better regulations. This includes reinstalling some of the post- Depression regulations that were removed by Congress during the last 20 years. (NYT, 10/14/14) This is just another example of weaknesses in our democratic experiment that needs fixing.

In the 1960s, Milton Friedman, a Nobel Laureate, was recognized as the Father of Monetarism when he linked steady growth in money and market-based pricing to economic growth. And now, in post-monetarism, the Great Recession wasn't supposed to happen based on that monetarism theory. This has led the Federal Reserve to take over much of our financial industry, with a central-banking approach of low interest rates and massive debt. These steps were necessary in trying to save us from a second Depression.

Obviously, this is still an experiment on moving from monetarism to Keynesian theory, the theory that guided us out of the Great Depression. See, history can teach, but we must use the evidence to correct the mistaken theory, especially in the United States where these history lessons lead us in the direction of centralized, expert policy making d for longer periods of time. It was the free market, private industry approach that led us into the mess we find ourselves in.

And now the NYT, 9/28/14// reported on *The Dismal Science*, just as what we have reported earlier on. Paul Krugman and Felix Salmon reviewed these two books:

1. *Seven Bad Ideas: How Mainstream Economists Have Damaged America and theWorld* by Jeff Madrick
2. *The Shifts and the Shocks: What We've Learned—and Still Need to Learn—from the Financial Crisis* by Martin Wolf.

Both books admit that mainstream economists did not believe the financial crisis could happen; their models predicted this could not happen. But worst, the economists were unable to respond. One main bad idea was Milton Friedman's case against government intervention not being needed because the free market can handle itself without regulations, etc. The economists then dismissed the empirical evidence already available prior to the financial system collapse. These learnings, plus political and professional reasons, only a few economists today admit, and that new models need to be developed; maybe using Piketty's big data, etc. We could go on and on, but this is not supposed to be a book on the science of economics. We only wanted to show how economics fits in with the other. Read these books plus the reviews; the mainstream scientists agree on all the examples. Each example has its own fear to overcome before we can move toward a successful future with minimum inequality/poverty, cheap energy, and a clean environment.

China is pushing to reform its economics. Hung Jing at the Singapore University said, "If you don't get rid of privilege groups, economic reform can't be carried out." An economist at Tsinghua University, Mr. Li Daokui, added: "Economic reform isn't in first gear." We believe the same thing could be said for the United States, Russia, and India (the big ones).

In his book *The China Model*, author Daniel Bell makes the case that China has a better governance model than Western democracies because voters are dumb, corruptible, gullible, and selfish and can be misled by politicians that are immoral and opportunistic with the assistance of commercial media that capitalize on ad hominem/feminnum and other profitable agendas. Bell writes that the China model is <u>moving</u> toward a political meritocracy that will prove to be a better governance system because Western democracies are too slow to recognize their weaknesses, much less correct them. Read Mr. Bell's book for a better and more complete explanation.

And now after our writings, Krugman acknowledges on 9/15/14, NYT, that economists got it wrong because economics is not a true science yet, and that "the fault lies not in our textbooks, but in ourselves." So let's all hope the Recession experience of the twenty-first century provides enough experience for dig data Science and knowledge for economics to become a "real science." Lord knows we need it.

With good economics, energy, and the other disciplines, capitalism and freedom can be saved not from external forces but from self-inflicted, "letting naked partisanship trump analysis" and not from factoring "faults" into a new analysis. Let's all hope economics can move forward with good science.

And now, we have a follow-up to Piketty's work via a U.S. economist, Pavlina R. Tcherneva from Bard College regarding her work on how the benefits from rising incomes and living standards have changed from the 1960s and 1970s up to today (2012). She not only fully supported Piketty's work, but she demonstrated that income inequality is getting worse every day. She displayed big data that show expansions in earlier periods (1940s onward); income went to most of the people, but in the most recent expansion (2009–2012), 98% of the income gains occurred in only the top 10% of earners. The bottom 90% achieved only 2% of the gains. In fact, in the last three years, incomes of the bottom 90% have fallen, whereas the top 10% captured 116% of income gains during that period (NYT, 9/27/14).

Now where is the WSJ after its opinion regarding Piketty's work based on "seemingly income inequalities?" Certainly not today, but I will keep looking.

23.0

Pandora's Promise

Pandora's Promise is a CNN special TV show directed by Mr. Stone. The special was shown on 11/7/13 at 9:00 p.m. The show demonstrated that nuclear power can stop gas emissions that cause climate change. Herein, we show that nuclear power is the only solution for cheap safe electric power, power the world needs. As we proceed forward into the twenty-first century, this power will:

1. Help all nations reduce poverty, especially developing nations
2. Save millions of lives from FF pollution
3. At least stabilize our climate, and with time, our climate can improve. Then with nuclear and other renewables, we will be releasing anticlimate CO_2 at a rate that our vegetation can convert back to O_2, Just like the world did prior to the FF era.
4. Help nations balance their imports and exports. The United States imported about $225 billion of FF in 2011. Some other countries have less of an imbalance, and some have more.
5. New reactor technologies will be capable of reusing the spent waste generated over the last 50 years.

Pandora's Promise got to the sequiturs via pollution concerns. Our research got there because of our concern about the impending energy crises due to the depletion of FF. For every kilowatt-hour saved with nuclear power, a life is saved, maybe two.

The new information I picked up from the TV special include these important points for our research:

1. The technology for fourth-generation reactors is further along than I knew about.
2. The *Chernobyl* accident wasn't as bad as I was led to believe. Only one person was killed directly and only 50 later.
3. 2 billion people are without any source of electricity
4. An iPhone uses as much electricity as a refrigerator.

5. Europe's electricity prices have gone up 17% for residents and 21% for businesses since Fukushima. This supports our research about the expensive wind energy in Germany and that France has lowered its energy cost 50% with nuclear plants.
6. FF have killed three billion people and 1.8 billion people have been saved because of the nuclear power we have today.
7. We will need three times the energy we use today by the end of the twenty-first century (supports our research).
8. Yucca Mountain, the storage place for nuclear spent waste has never opened because of politics. This is a better place than where the waste is now, even though the 70,000 tons we have now is not an environmental issue. Again, once the Obama administration releases funding for the IFR program, the waste can be used for fuel, just like the French does, a much better approach than food to fuel.

Other points of interest from the CNN special:

1. Renewables amount to only 2% of world power (we also reported this earlier), with demand growing at 2% per year. This was in response to EV claiming renewables can supply our growth needs, as well as replacements for FF at competitive prices. This is the entire motive for the EV movement, along with the killing issue. Ralph Nader is the EV leader.
2. Replacing FF with nuclear will save a total of 7 mm lives. With 3mm killed already because of the FF emissions.
3. EA politicized the nuclear plant built in New York. The EA campaign was financed with King Coal and Big Oil money, and some of this money came from taxpayers via the alcohol fuel tax ($7.0 billion/year) rebate.
4. Because of EA protests with lies/fear, the New York nuclear plant was never started up; just like the Farmers Insurance ad, "knowledge can provide a plan for the future."
5. Nuclear power can bring countries out of poverty. We didn't go that far. Our plan fully supports this claim.
6. Even the kill rate for solar is worse than nuclear, given that there are 2.0 deaths per tWh for solar vs. 0.5 deaths per tWh for nuclear providing millions of times more power than the sun. And just today, 5/15/14, 300 coal miners killed in Turkey,

two in West Virginia. *Pandora* doesn't say anything about the nuclear comparison with coal; it's obvious with three million compared with 50 deaths.
7. The new-generation IFR will discharge waste with a half-life of 800 years vs. 10,000 years for present reactors. President Obama, "listen to this."
8. One could drink all the water from an NPP and get less radiation than eating one banana.
9. Background radiation is higher in most places than inside an NPP. Background radiation ranges from very low to about 0.2 rem in Brazil. The importance here is that world organizations have studied the cancer rate across the spectrum and have found no correlation. (Reread Chapter 20, I think it's there! Maybe 19.)

Crossfire regarding *Pandora's Promise*, continued

CNN organized a panel to discuss the special. The panel consisted of four people:

1. Newt Gingrich,
2. Ralph Nader (EV)
3. Lady (EA)
4. An ex-EV turned pronuclear after studying and discussing with experts

The discussion doesn't need to be summarized because the points were brought out in the special. It was the first time Newt was the only sensible one. Now I'm mad. When I was a young man, I became a Jane Fonda antinuke because of all the lies of the EV mendacious propaganda in the early 1960s when I was 18–20 years old and was too busy studying to research the subject. So today, I plan to use all my energy to expose these "lies" and promote nuclear.

I use the word "lies" which is strong because the truth is so readily available. Any thinking person should know better. Ralph Nader's whole life is built around these issues. So he would destroy his livelihood if he admits the truth. And in our opinion, a person that won't change in the face of so much information isn't much of a person.

One last observation: The special had old clips of Jane Fonda screaming the "lies." (Even at that time, they should have known better.)

I will give Jane credit though; at least, she is not on the frontlines today. Is she still alive? Maybe the pollution killed her.
Philippines — Post *Pandora's Promise*

The climate change inequities were brought to the forefront with typhoon Haiyan's destruction and death toll in the Philippines (NYT, 11/17/13) reaching 6,500 casualties.

Poorer countries are demanding that countries that contributed to the carbon problem help out poorer countries that didn't contribute much to the problem. This was highlighted in the IPCC report.

Now, what are we going to do if CO_2 levels double because we are burning up all the FF? Does anybody want to pay up after being duped into thinking climate change wasn't a human activity, and you continued having a big CO_2 footprint? Quite frankly, as a scientist/engineer, I don't understand why this issue is controversial, given these reasons:

1. We can calculate how much CO_2 is released by all sources, and it doesn't go anywhere except into the atmosphere.
2. Scientist have known forever that CO_2 provides the blanket to keep us warm. The CO_2 released from FF is far and away the predominant source of anticlimate emissions.

And now, before we could finish this book, typhoon *Hagupit* (Ruby) plowed through the Philippines, dumping 200 milliliters of rain as it plowed through, killing many and forcing over one million people (10% of the country's population) from their homes. Sea waves were over 20 feet high. The last time I checked 2+2 = 4 (without a calculator).

Just this year we have witnessed:

1. The largest tornado at 1.3 miles wide in Oklahoma, Nebraska, etc.
2. And now the largest hurricane with winds up to 230 miles per hour. No one wants to speculate on how bad the weather will be with CO_2 at 800 ppm.

This is a world community problem, and it will take the world to solve the problem. The IPCC has laid out the evidence before us. God gave us the brains and the evolution process, so now, it's up to us to use them.

This chapter will complete the book. Last night, I was watching an MSNBC program with immigration activists as guests. They were discussing Martin Luther King's book, *Why We Can't Wait*, and the activists said they were inspired by the book. It reminded me of my connection with Dr. King. I marched with Dr. King from Selma to Montgomery, Alabama. Auburn is close to Montgomery. I was late, so I marched behind the main group. I didn't meet Dr. King personally. I was thinking, as the activist talked, about the Dr. King's book. It would be nice to have a concise statement for our nuclear renaissance, so I have coined this motto: "A risk path going forward is a better choice than a safe path going backwards."

Added to this is our label 2C power for the world integrated with 2C lower temperatures. And this week, I went to see the movie *Selma*. After the first scene, I started crying, and as I reviewed our book, I'm still crying. My next book will be titled "A White Nigger Living in Alabama (1965)."

Another eventual happening for me was when a couple of friends of mine and I went into enemy territory at Tuscaloosa, Alabama, home of the University of Alabama. We attended a speech by Robert Kennedy, the last he gave before going to California where he was assassinated. I cried all night. What I remember most about the speech was his warm handshake after the speech. The reason I was so impressed was because every time I give an important speech to a group, my hands get ice-cold. I'm impressed with Obama's speeches because he is so good at delivering the message. Someday, I hope to shake his hand also.

A third event I attended was a speech by President Reagan in 1992 at the International Construction Round Table Conference in Monterey, California. I was president of the local chapter in Ohio. This was President Reagan's last speech before calling it quits at 92 years old. What my family and I remember about this speech was the humorous stories he could still deliver. Shortly after this event, it was announced that he had Alzheimer's. I hope when I get to 92, I can be as eloquent and humorous as he was.

Just one more story before we leave. This week, I asked a homeless man to go to a thanksgiving dinner with me at the corner of Crenshaw/Adams in Los Angeles. He stood me up. I guess he was too lazy to walk to the corner. Restauracion, a Christian group, set up a free dinner for Thanksgiving. They were giving away old clothes with the dinner. I ended up eating with 200 other hungry souls.

A pretty young lady escorted me to my table; two young boys brought me my coffee/water and turkey meal with pumpkin pie. I did not have to stand in line. It was the best Thanksgiving dinner I ever had, with the exception of some with my second ex-wife. That's what this country is all about. The poor sometimes enjoying life better than the rich. Because of this experience, I have committed to give 25% of my book's profits to Restauracion.

Until I am accepted for the renaissance position, I will be acting as if I already have the position. Since I stopped short on the book, I will be issuing updates. I don't have a "blog," but maybe I will get one for these updates.

    stopthefuelishness@yahoo.com
    Date: 01/01/2014 (updated 4/15/15)

And just today in the NYT (8-14-15), we read about two events in China that support the claim that FF are killing millions of people every year.

First, the peer reviewed science journal PLOS ONE reports that a study completed this year proves that 1.6 million people die each year from FF emissions related to the exposure to fine particulate matter in the atmosphere. These findings are based on data extracted hourly from 1,500 stations throughout China. The authors work for Berkeley Earth, a research organization based in Berkeley, California. They used statistical techniques such as Big Dada to quantify the results. They also extracted information and algorithms from the framework used by the World Health Organization (WHO) for projecting death rates from five diseases known to be associated with fine particulate matter. The raw data and algorithms have been published so that others can review their work.

Second is the FF chemical explosion in Tianjin, China, that killed hundreds and evacuated thousands. The atmosphere is filled with toxic chemicals such as sodium cyanide, which led to even more deaths. The mushroom cloud and fires were worse than the hypothetical cloud shown in the movie *The China Syndrome*. And the shaking of the earth was worse than the latest tsunami that occurred in Fukushima, Japan.

It appears China has adopted a conservatives' approach to capitalism, that is, profits without safety or environmental regulations.

After all, implementing these regulations reduce profitability of the venture. And many times, the owners can pocket the extra money before a disaster occurs. This happened in the United States with Hurricane Katrina. Politicians think they can stay in office longer by not spending money and increasing taxes, as recommended by engineers. As a chemical engineer, I read about the FF chemical plant in China regarding storing 1,300 tons of ammonium nitrate and potassium nitrate alongside 500 tons of magnesium powder.

Calcium carbide, sodium hydrosulfide, and toluene diisocyanate (both extremely toxic chemicals) were stacked in containers three containers high (this by itself is a severe violation of safety regulations). And to have all these chemicals stored together is like having a ready-made toxic bomb waiting for a minor spark to set it off. No engineer would ever work in such a powder keg situation; however, it happens all the time because owners will fire those that complain.
Ending

Mission Accomplished

You already know what I tell my friends when it's our turn to act. If you forgot, go back to Go. As a great man, Nelson Mandela, commented "from dust to dust," and I add to it "from poor as a child."

Back to poor as a senior: "What a journey! Thank you, my loving Jesus. I wouldn't have it any other way, except maybe a little less inequality."

24.

Closing Statement

And now, after all this work, we find ourselves with the only answer for our threefold problems:
1. To save our environment
2. To have cheap energy for centuries of generations
3. To power the poor out of poverty

The answer is *nuclear power*. Leading a worldwide sustainable energy policy.

Even when I tell my friends (some are elite, educated people) about our research, they will turn up their noses and say, "No, not nuclear." We don't have any illusions that "we don't have a herculean task ahead of us." We do have the guiding light of the four countries that are mostly on the risk path already. So our task is achievable; just with a little monarchy (not too much) and a renaissance man to lead the world collective.

The United States, a democratic experiment, has shown to the world what a collective of poor people can achieve when they are free and have a little reserve to use their collective talents and to improve the status of all our brothers and sisters. So now let's solve the democratic weaknesses and move forward with the power of nuclear and save our atmosphere at the same time.

Pandora's Promise has visually shown us that nuclear power is the only means to save the environment, and now with our work, we not only support NPP, we have demonstrated by numbers and words that nuclear power saves the world economies and energy for many generations. The primary hurdles to overcome are education and fear. The new energy book by Bryce that we referenced above reaches the same "basic conclusion regarding the use of FF until a nuclear program is built-out." We detail a comprehensive plan and provide numbers to support our findings. The one difference is that we will not run out of

FF because the marketplace will balance supply and demand. (Bryce is funded by the conservative group, "Manhattan Institute.") Demand will balance supply with higher energy prices.

Our comment is that the caveat regarding a 2% growth will take over and cause higher prices and reduce the growth to less than 2%. In the meantime, if energy prices increase over the next 50 years, like the prices have over the past 50 years, gasoline (oil) will cost $100 per gallon, natural gas will cost $100.00 per million Btu, and electricity will cost $1.00 per kWh. These higher prices appear to be unreasonable, but these are the same increases that have occurred over the last 50 years. We have made reference to the price of gasoline in the 1960s going from 26 cents per gallon, a 2,000% increase. We say many more people will not have energy at these prices. The last few pounds of FF will only be affordable for the top 1% of the population.

New information, facts and opinions are not covered herein:

The United States has the cheapest power at 12 cents/kWh vs. 26 cents in Europe, 24 cents in Japan, and 24 cents in Germany. And the reason is that the United States currently generates more power from nuclear than any other Western country in the world. But as explained herein, the United States is losing ground fast to at least four other world powers on nuclear power, with no plan to change paths. We already read about China developing the best, cheapest, and safest nuclear technology with its engineers having access to both Eastern and Western technologies and prototypes and remember the best comes from the West.

The NYT, 9/8/14. *Let the River Run Wild* by Waldman et al. tells about the Susquehanna River being one of America's greatest treasures, but we destroyed its natural beauty and biodiversity with four dams. And we need to destroy these dams now; "the dams are gradually destroying themselves with alluvium build-up that is a function of all dams everywhere." Waldman proposes that solar panels be installed over the abandoned lake beds, etc. to replace the 575 megawatts the dams generate. Waldman is on the right track, but solar panels in New England is not a good solution. It's like wind generation out in the oceans. Only the wealthy would be able to afford the electricity price. And the United States would become another Germany. A better alternative is to build a couple on NPP and get rid of all the dams. See, our collective can join up with a few knowledgeable EA, but not with EV although: because of their lies/fear and criminal activities like the report in WSJ (12/29/14) on EV damage to the geoglyph of Peru in the "Time for change" message written by Greenpeace near the ancient Peruvian site. By the way, these solar panels will blanket the entire valley, destroying the biodiversity. Some EA will be joining our collective.

Speaking of damming rivers, China has more dams than the rest of the world combined. And it has plans to double hydropower by damming the Tibet rivers over the next decade. Tibet has seven major rivers that spread across the world's largest delta across Asia. China is doing this because it needs to reduce FF power, until China's nuclear program is built. We believe China is shortsighted to do this even for a short period of time. A better plan would be to expedite the NPP and let the rivers run free. See, we can be on the side as EA and freedom-lovers also. The health of these rivers is important not only for China but also the health of the world's ecosystems.

And Africa has a choice to make soon: Hydropower or Nuclear.

Either one will provide 2C power.

One last addition. It's taking longer than I thought to get this book published; and the evidence keeps coming in. A committee of the American Association for the Advancement of Science, the world's largest general scientific society, released its report on 3/19/14. The conclusion:

The effects of global climate change are already being felt, warning "that the effects of human emissions of anticlimate gases and the ultimate consequences will be dire, and the window to surcease it, is closing very fast. The evidence is overwhelming. Levels of anticlimate gases in the atmosphere are rising alarmingly fast."

This is just the latest report that continues to support and warn the world that unless we change course, the world will suffer the consequences. The committee's last finding: Scientists bear some of the responsibility for the public not believing the scientific data because, in their well-meaning efforts to convey all the nuances and complexities, the basic message is obscured regarding the risks going forward. They hope this latest report will clear up the clutter, and the world will hear the message. And that's why I'm in such a hurry to get this book published in order to support the efforts of all these great scientists and engineers of the world, and to put to bed Matt, Brit, Aye and a couple of others.

Along with the scientists' report on anticlimate gasses, another group of scientists from the Lawrence Livermore National Laboratory reported that they are making headway with the science of fusion energy, another source of endless energy once the science is developed. The latest hopeful report for the world from these scientists is that by using Livermore's giant lasers, they were able to fuse hydrogen atoms and produce flashes of energy, like miniature hydrogen bombs. The amount of energy was small, but it was five times what had been achieved before. So this work on endless energy for the world is continuing, and our scientists will be successful, as long as the public and politicians stay behind the effort and believe and act on the results. This will take funding from the government via international organizations. The evolution is continuing to work.

São Paulo, South America's biggest and wealthiest city with a population of 20 million, is running out of water, with key reservoirs already dried up because of the drought that has lasted more than 50 years. And we say, what about the U.S. West where key lakes are down over 80% (NYT, 11/5/14)? And then today, it rained two inches. Since we understand sense perception, our conclusions will not be changed.

Nuclear Transportation
Now we answer the question.

LAT, 12/27/14: The LA Metro System is testing a 60-foot electric-powered bus that's capable of transporting 120 people over 170 miles per charge. The batteries (lithium) are supplied in a pack, eight total, four to a side. The bus is a prototype built in Lancaster, California by BYD Motors Inc., and I bet you can guess where BYD headquarters are. That's right! China. BYD has been manufacturing buses for some 11 years. LA Metro has specified to BYD what is needed in Los Angeles to be successful—efficiency that will satisfy the public. The LA prototype is an accordion-like articulation 60 feet long. Orders for BYD-esques buses are not new to the United States, particularly in San Antonio, Pomona, and Washington (the tricities). These cities already have 40-foot long buses. Los Angeles County has put in an order for forty of the 40-foot buses and will use these until the 60-foot accordions meet all the LA specifications.

LA Metro is already a world leader with the largest natural gas fleet of buses in the world. Now LA Metro and China are learning from each other. James Holts, sales manager for BYD, stated: "It's an opportunity for there to be a renaissance in public transportation." Since China is already on the nuclear risk path for clean 2C electricity, it is also converting its trains to electricity. Trains are easier because they are easier to connect directly into the electric grid.

So now we have answered the question. How about cars? Not needed once you ride in an LA Metro electric bus or a simulacrum therewith. And better yet, build an electric grid system to connect buses directly to the electric grid. This will "practically" eliminate the need for cars and batteries. Dayton, Ohio; New Orleans; and San Francisco are cities that already have partial electric grid systems like this, leftovers from the 1920s era becoming outdated before realizing the carbon issue and cheaper FF. Again, cream rises to the top. As long as the Supreme Court doesn't interfere with the evolution process. Also, give credit the governments of these cities for the few cars/trucks needed locally; these should be electricity-operated without batteries.

You can believe me because I own a 150 Ford here in Los Angeles, and I never use it for transportation, except when I travel to the Philippines. The Ford van is my home here in Los Angeles. When the oceans rise 23 feet, I will have to abandon the buses and use the van for transportation and as my home.

Once progressive cities discover that LA Metro is chauffeuring people around, point-to-point using fuel that cost 2C per million BTU - about 2C per mile, with customers loving the experience - all the oligarchs will jump on board also. Extra: LA buses carry all passengers (no discrimination), and blacks, Asians, and Latinos don't have to move to the back so that whites get the preferred seats. The LA buses are equipped to carry the disabled in wheelchairs and bicycles for the adventurous and poorer people that can't afford the total fare.

Yesterday I witnessed a totally blind person with her dog riding the bus from point-to-point. All buses have recordings that call out the stops along the route. This is the work of evolution in progress. And China and Los Angeles are the leaders on transportation. And we haven't even discussed how rail and bullet trains fit into the mix. Give

Los Angeles a couple of years then visit and find out. Villaraigosa was one of the best leaders of Los Angeles, getting a tax increase passed for buses and the bullet electric trains from San Francisco to Los Angeles, connecting with the orange line. You can even visit now and witness the progress at work.

WSJ 10/7/14

An opinion by Mr. Leonard representing the WSJ came as close to recommending nuclear power as we can expect. Mr. Leonard CEO of Global Environment fund recommends a two-part program financed by a carbon tax to support R & D and prototype projects, including nuclear power. Not much, but at least a start. Now we can build on this start.

And now, literally after completing the last task herein regarding a proposed transportation which could go nuclear with electric, LA Metro buses tethered to the electric grid. It came to me that if buses and trains can be tethered, what about cars and trucks? This has never been done before. An engineer and project manager doesn't stop here. We have the technology, engineers, project managers, designers, and all other help (jobs for the unemployed), plus the resources. Why not include local cars and trucks connected to the grid if we cannot give up private transportation?

Simply put, we envision vehicles plugged into a socket on a raceway located to the right side or overhead. These tethered vehicles would travel to an intersection where the control system will shift the socket to the programmed new route. This new system would not have the flexibility to drive 100 mph or turn into oncoming or side traffic, causing an accident.

The programmed route would be for your vehicle only. This will be the largest project any country or government has ever undertaken. The project would start by designing and engineering the system for a large city like Los Angeles or Beijing, then moving on to the countryside, then moving on to the whole country. Prototypes would be developed for each system and the proof-of-concept tested and approved before moving on.

China, being as progressive as it is, may already have started on a project like this or a simulacrum. The next step is to develop the PFDs (process flow sheet diagrams then develop the PIDs, followed by a complete listing of equipment and instruments. This project should be undertaken by the UN, using the new council (UNSEC). The engineering package can be developed for one country then shared with the world.

Our ancestors built monuments and other huge projects using nothing but brawn and guts. The pyramids throughout the world are examples. The Seven Wonders of the World include a few. And here in the United States, we have the Hoover Dam, the space mission, and the Watts nuclear power plants. We live in an era of great capabilities to engineer huge projects; these projects couldn't even be dreamed about, prior to evolving to the position we occupy today. It will take a huge collective of humans all over the world sharing knowledge with each other. And everybody on these projects will be paid equally, without any discrimination, except maybe an Ayn, a Norquist, or a Nader.

And nuclear is only the starting block. The resources and supplies of the world, plus the ongoing research on thorium fusion and the ITER reactor. Actions like these will make it extremely difficult for the oligarchs to control the power of the world. Like they do with FF resources. The exception here is a big country cornering the world's supply of uranium (Russia) with America's help.

One final point: Don't let a project of this size scare you. I'm telling you, no matter the investment, the returns will be more than 10% per year for 25,000 years, plus all the deaths and injuries we saved. No debt is large enough to stop a project with returns this large and with a side benefit of saving our atmosphere

A point about the 25,000 year sustainability: The 25,000-year prediction has an important caveat. It represents consumption with 2% growth. If or when countries start consuming more than this caveat, the prediction will change. What's really needed is a graph on world percentage conversion to nuclear versus sustainable years of nuclear supply.

I could do this, but I don't have the time, energy, or resources. The new sustainability council should take on this task and include it in its annual report, along with recommendations for modifying energy programs based on results. I'm sure this will show that the West has no business selling nuclear supplies to Russia.

Our ancestors passed down to us the saying about will. "Where there's a will, there's always a way." Notice how we slightly embellished the saying. That's all folks, until next time!

Final Thought

The epiphany came to me in the middle of the night, September 20, 2015, when I woke up in a cold sweat, thinking about chain reactions. No, don't go there. I was thinking about the chain reactions of climate change. Bear with me for a bit longer with some chemistry related to the matter.

A rule of thumb for reactions: "For every 10-degree increase in temperature, reaction rates double." Also, the absorption of a minor chemical in the body of a major chemical is inversely proportional to an increase in temperatures.

<u>First</u> chain reaction:

The primary reaction for the burning of FF is
$C + O_2 = CO_2$.

For every pound of C, two pounds of O_2 (oxygen) is taken out of the atmosphere, and three pounds of CO_2 are created.

But worse than that, the volume of C creates twenty times the volume of CO_2 gas (without a calculator). This volume of CO_2 gas displaces the same volume of O_2. Some of the secondary and tertiary reactions are these:

1. The warmer climate creates more fires with a double-barrel effect.
   a. The "green" and dry vegetation C is burnt to CO_2.
   b. The vegetation is not available for photosynthesis of CO_2 back to O_2.

2. The ice melts, releasing the "trapped CO_2 stored since the beginning of time. Go through the exercise for droughts also.

<u>Now</u> onto the absorption issue. Scientists have always known that all water bodies are a sink for extra CO_2 in the atmosphere. In the past, when waters were colder, the CO_2 levels were X. As the temperature of these waters get higher, the quantity of CO_2 these waters can hold is some fraction of X. The other fraction of X is additional CO_2 released

to the atmosphere. We don't hear much about this additional release of CO_2 for two reasons:
  a) It's very difficult to calculate.
  b) It probably doesn't matter because of this reason:

Our prediction of a CO_2 stabilization value at 800 ppm is based <u>only</u> on the primary reactions without considering the secondary reactions. Today the bees and birds are dying off at a CO_2 level of 400 ppm. So maybe humans will be dying off at a CO_2 level less than 800 ppm. So the worst of the worst predictions are not needed.

The End

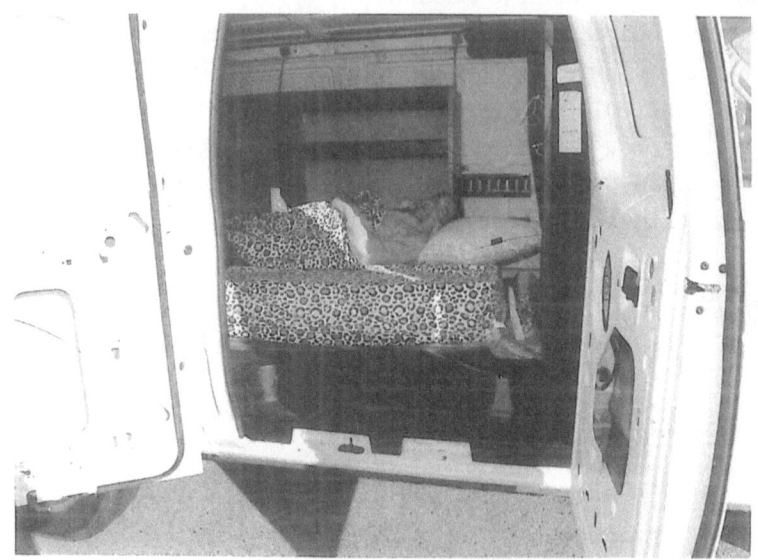

Home Sweet Home

Office

A good neighbor to Gardena

A good University for Engineers/Veterinarians
Ex-Transportation

Ohio PES 5011-R (Rev. 10/2013)
State Board of Registration for
Professional Engineers and Surveyors
50 West Broad Street, Suite 1820
Columbus, Ohio 43215-5905

**2014-2015 ONLINE RENEWAL
INFORMATION ENCLOSED**

The image that was added yester year -
Ray goes afer Truth - Honesty - Knowledge -
Opening statement - First page.

www.ingramcontent.com/pod-product-compliance
Lightning Source LLC
Chambersburg PA
CBHW030928180526
45163CB00002B/492